Waterflooding: Chemistry

Society of Petroleum Engineers

Richardson, Texas, USA

Disclaimer

This book was prepared by members of the Society of Petroleum Engineers and their well-qualified colleagues from material published in the recognized technical literature and from their own individual experience and expertise. While the material presented is believed to be based on sound technical knowledge, neither the Society of Petroleum Engineers nor any of the authors or editors herein provide a warranty either expressed or implied in its application. Correspondingly, the discussion of materials, methods, or techniques that may be covered by letters patents implies no freedom to use such materials, methods, or techniques without permission through appropriate licensing. Nothing described within this book should be construed to lessen the need to apply sound engineering judgment nor to carefully apply accepted engineering practices in the design, implementation, or application of the techniques described herein.

ISBN: 978-1-61399-794-9 [Print]
ISBN: 978-1-61399-795-6 [Mobi (Amazon)]
ISBN: 978-1-61399-796-3 [Epub (iTunes)]
ISBN: 978-1-61399-797-0 [WebPDF (ADE)]

10 9 8 7 6 5 4 3 2 1

Society of Petroleum Engineers
222 Palisades Creek Drive
Richardson, TX 75080-2040 USA

http://store.spe.org
service@spe.org
1.972.952.9393

I would like to thank Royal Dutch Shell for the opportunity to perform such a fascinating role, which enabled me to build a wide skill set in waterflooding and to address waterflood issues in fields all over the world. My thanks also go to the many colleagues and associates who, over the years, have supported, steered, and guided me. Without them, I would know nothing.
—Dave Chappell

Table of Contents

WATERFLOODING: CHEMISTRY

Dave Chappell

Dave Chappell has spent his career working on waterflood developments and operations in Brunei, Oman, Thailand, and Australia. In 2003, he became one of the founding members of Shell's central waterflood team tasked with improving waterflood performance across the entire Shell waterflood portfolio, based in The Hague, The Netherlands. He went on to manage that group from 2008 until his retirement in 2018. Since then, he has worked as an independent consultant in the waterflood arena.

1. Background

Historically, waterflood has been viewed as largely the responsibility of the reservoir engineer. That was primarily because waterflood considerations tended to be heavily focused on the displacement process and the associated recovery impacts. While those are critically important to waterflood success, it is now much more widely recognized that waterflood is a process with many different moving parts, and it therefore requires the input of a wide range of disciplines, each of which needs to interface effectively with the others, to deliver a fully optimized project.

Waterflood remains by far the most widely used process that uses an external energy source to improve recovery. Furthermore, it has been successfully used for more than a century. The theoretical basis for waterflood displacement has been understood for quite some time and is well-covered in the available literature (Willhite 1986; Warner 2015). Despite the focus on this aspect of waterflood, it is only in recent years that there has been any detailed inspection of the critical success factors across the full range of disciplines involved. Furthermore, not all the factors have been particularly well-documented. This book is one in a series of publications that aims to redress some of those shortcomings by looking at a range of factors influencing success in waterflood design and operation.

2. Introduction

It is becoming increasingly recognized that chemistry issues can have a significant impact on the way a waterflood project progresses. This topic is therefore the first to be covered in a series of books addressing various factors associated with waterflooding. These issues might primarily affect the operating costs associated with a project, but they can also impact recovery. For example, in fields where reservoir souring has occurred, there can be constraints related to the hydrogen sulfide (H_2S) content of gas export volumes. In extreme cases, these constraints could require wells producing high volumes of H_2S to be shut in, and production will, at best, be deferred, but there might also be a negative impact on ultimate recovery. Similarly, high-H_2S-producing wells might need to be shut in because of well-integrity concerns associated with the corrosion impacts on well materials. In such cases, it is highly likely that stranded oil volumes cannot then be produced because the remaining volumes do not facilitate economic repairs, so this factor suggests a very high likelihood of an overall recovery impact.

This is just one possible impact that chemistry can have on waterflood performance. In addition to reservoir souring, chemistry impacts in waterfloods include water/rock interactions that not only can have negative consequences for injectivity but also might affect recovery through changes to rock wettability (low-salinity flooding), scaling problems, wax and asphaltene problems, and hydrate formation. Consequently, a range of chemistry-related issues can materially influence waterfloods. These issues will be explored in more detail in this book.

3. Reservoir Souring

Reservoir souring is the phenomenon in which producing wells begin to produce H_2S even though this gas was not present in the reservoir before production began. This problem is commonly encountered in waterfloods, particularly in those where seawater is used as the water-injection source. It is thought that approximately 80% of seawater floods have encountered a degree of reservoir souring at some time during field life.

3.1 Negative Effects of H_2S. The issue of reservoir souring has attracted considerable interest over the years because there are significant issues arising from the presence of H_2S gas, including health, safety, and environment (HSE) concerns; corrosion; and gas sales contractual limits.

3.1.1 Health, Safety, and Environment Exposure. Because H_2S gas is toxic, operational procedures are needed to keep staff safe. At low concentrations, H_2S has a characteristic smell of rotten eggs. However, as concentrations rise to between 100 to 150 ppm, the sense of smell becomes paralyzed very quickly, and therefore the gas can no longer be smelled (olfactory desensitization). Exposure to higher levels of the gas can deaden the sense of smell instantly. At concentrations above 500 ppm, a person can collapse within 5 minutes, leading to death within 30 to 60 minutes. The effects depend on how much H_2S is breathed in, and for how long, but exposure to very high concentrations can kill a person quickly. The consequences of exposure to H_2S in humans is summarized in **Table 1.**

Table 1—Effects of H_2S gas and the response in humans.

H_2S Concentration (ppm)	H_2S Effects and Human Response
0.025	Detectable smell
10	Maximum exposure limit of 8 hours. Sore eyes will develop.
15	Maximum exposure limit of 15 minutes
20 to 30	Intense offensive odor
50	Rapid conjunctival irritation
100	Olfactory fatigue. Sense of smell paralyzed in a few minutes. Stinging sensation in eyes and throat.
500	Exposure at these levels expected to induce collapse within 5 minutes. Respiratory paralysis in 30 to 45 minutes leading to death unless external resuscitation applied.
1,000	Rendered unconscious very quickly. Permanent brain damage can result unless rescued promptly. Imminent death.

If leaks occur in a production process, the H_2S will tend to disperse, which helps to mitigate the risks. However, this might not always occur because H_2S gas is heavier than air, and consequently, it tends to concentrate and accumulate in any lower-lying, stagnant areas. This can easily lead to high H_2S concentrations accumulating locally in locations such as well cellars, for instance. This can make any work in confined areas particularly dangerous in fields that produce H_2S.

Finding an unconscious person when H_2S can be smelled is a particularly dangerous scenario because the concentrations might be significantly higher at the location of the body, and they might be high enough to cause an almost-instantaneous loss of consciousness. It is vitally important, therefore, that an unconscious person not be approached in such a scenario without taking appropriate protective-equipment precautions.

3.1.2 Corrosion. H_2S can cause severe corrosion problems. The generalized corrosion rates associated with H_2S can be quite low, but locally, there can be severe corrosion problems associated with stress corrosion cracking, hydrogen embrittlement, or hydrogen-induced cracking processes. In the case of reservoir souring, these problems can be magnified because the H_2S was not initially present, and, as a consequence, it is possible that the materials used will not be compatible with the subsequent concentrations of H_2S encountered. This can lead to catastrophic corrosion failure and can necessitate expensive materials changes to wells, in particular.

The corrosion of metals in the presence of H_2S is well-known (Cord-Ruwisch et al. 1987). One potential mechanism takes place by means of cathodic depolarization, described as follows.

Anodic reaction:

$$4Fe \rightarrow 4Fe^{2+} + 8e^-. \dots\dots\dots\dots\dots\dots\dots\dots\dots\dots\dots(1)$$

Water dissociation:

$$8H_2O \rightarrow 8H^+ + 8OH^-. \dots\dots\dots(2)$$

Cathodic reaction:

$$8H^+ + 8e^- \rightarrow 8H^\bullet \rightarrow 4H_2. \dots\dots\dots(3)$$

Hydrogen oxidation:

$$SO_4^{2-} + 4H_2 \rightarrow H_2S + 2H_2O + 2OH^-. \dots\dots\dots(4)$$

Precipitation:

$$Fe^{2+} + H_2S \rightarrow FeS + 2H^+. \dots\dots\dots(5)$$

Total reaction:

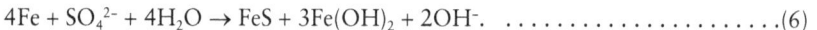

$$4Fe + SO_4^{2-} + 4H_2O \rightarrow FeS + 3Fe(OH)_2 + 2OH^-. \dots\dots\dots(6)$$

The cathodic reaction described in Eq. 3 shows the formation of hydrogen molecules. However, this reaction proceeds by means of hydrogen atoms (also called hydrogen radicals) that are extremely reactive, and normally, they recombine very quickly with another hydrogen atom to form a hydrogen molecule (H_2). However, H_2S tends to poison and retard this recombination reaction, and as a result, the hydrogen radicals persist much longer than they otherwise would. It is possible for hydrogen to penetrate the lattice structure of the metal, and because a hydrogen radical is much smaller than a hydrogen molecule, the rate of penetration is much greater for a radical than it is for a molecule.

The radicals can diffuse through the grains and other lattice defects. Eventually, the recombination to a hydrogen molecule will occur, and because the hydrogen molecule is larger, it can weaken the metal at the grain boundary. Thus, stress corrosion cracking results both from the stresses induced by the internal pressure as a result of the recombination of hydrogen on specific sites in the microstructure and from a hydrogen embrittling effect that weakens the cohesive forces within the metal.

Badrak (2018) suggests that when the partial pressure of H_2S exceeds 0.05 psi, sour service materials should be used to protect against the danger of stress corrosion problems. Because of the large number of floods using water sources containing sulfate, using these materials could be a sensible precaution when waterflooding in such environments. It has been estimated that sour service metallurgy has a cost premium of 1 to 2% of the project cost (Al-Rasheedi et al. 1999), although this premium varies at different times. This might be considered an appropriate insurance when weighed against the much more significant costs associated with retrofitting such metallurgy.

3.1.3 Gas Sales. Because of the corrosive nature of H_2S, gas sales contracts often set a maximum-allowable H_2S content—typically in the range of 4 to 6 ppm.

Thus, when reservoir souring occurs, these contractual limits can potentially be breached. This would demand that either high-H_2S-concentration wells be shut in or expensive H_2S-removal technologies be used. This can result in very high levels of deferment in fields that sour and where the gas is sold or in a high operating cost for its removal.

3.2 What Causes Reservoir Souring in Waterfloods? There are a number of different potential mechanisms whereby H_2S could theoretically be generated in a reservoir. These include thermal-generation methods such as the thermochemical reduction of sulfate by hydrocarbons or the thermal hydrolysis of thiophene compounds (organic sulfur-bearing molecules). However, the temperatures required to initiate such processes are high enough that they can be effectively discounted as the mechanism for reservoir souring in waterflood environments. The injection of water will usually result in local temperature reductions rather than increases, so if the temperature is high enough to induce thermal H_2S-generation processes, they will have already occurred before the waterflood began.

The connate water has some capacity to absorb and store H_2S. If the reservoir pressure decreases during production, the solubility of H_2S in the water will decrease, liberating H_2S from the water and moving it into the produced reservoir fluids. This was identified as a mechanism responsible for increased H_2S production in the Caroline Field in Canada (Seto and Beliveau 2000). However, this mechanism requires that H_2S be present in the reservoir fluids at the outset, which is not the case in the vast majority of fields where reservoir souring is encountered. Furthermore, it also requires that the reservoir pressure be reduced before it manifests, and in most modern waterfloods, the reservoir pressure is usually maintained. Nevertheless, many waterflood projects still see increasing H_2S levels produced over time.

With the benefit of hindsight, it is fairly surprising that the mechanisms behind reservoir souring in waterfloods remained somewhat of a mystery for so long. However, it eventually became evident that reservoir souring in waterfloods was attributable to the activity of sulfate-reducing bacteria (SRB) (Herbert et al. 1985). This is now universally recognized, but if there is any doubt regarding the source of the produced H_2S in any given field, it can be confirmed by performing a sulfur isotope ratio on the produced H_2S. Sulfur has two main isotopes—^{32}S and ^{34}S—and bacteria prefer to use the slightly lighter ^{32}S isotope. The preferred consumption of the ^{32}S isotope consequently results in a slight enrichment of the ^{34}S isotope in the sulfate, although, as can be seen in **Fig. 1**, the changes in the isotope ratio are small.

This analysis requires a "standard" sulfur isotope ratio, and the one found in the troilite nodules (iron sulfide) within the Canyon Diablo meteorite is commonly used. Because the fractional variations are modest, the differences are typically expressed as parts per thousand:

$$\delta^{34}S \ (\text{in } \text{\textperthousand}) = \left[\frac{^{34}S/^{32}S \ (\text{sample})}{^{34}S/^{32}S \ (\text{standard})} - 1 \right] \times 1,000. \quad \dots\dots\dots\dots\dots\dots\dots\dots\dots \ (7)$$

Perhaps one reason that acceptance of the mechanism for the phenomenon took some time was the knowledge that most bacterial species tend to thrive in modest temperatures—up to 30 to 40°C for mesophilic SRB—whereas reservoir temperatures are typically appreciably higher than this. However, there are a large number of species (thermophilic SRB) capable of thriving at much higher temperatures.

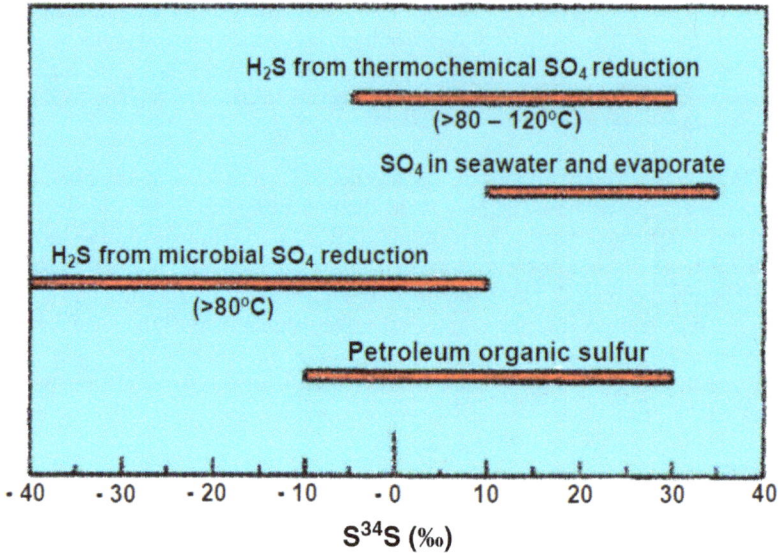

Fig. 1—Variation in ^{34}S isotope ratios in biological and nonbiological sulfate (SO_4) reductions (after Herbert et al. 1985).

Indeed, SRB can be considered almost ubiquitous. They are a very diverse and versatile class of bacteria (Muyzer and Stams 2008), and although in the context of this subject we will be looking at specific metabolic activity, they can be very versatile in their use of various electron acceptors and donors. Furthermore, they can thrive in an enormously diverse range of environmental conditions.

There has been some debate as to whether their spores can survive in a dormant state over geological time in an oil field while awaiting suitable conditions for growth. However, that consideration becomes somewhat hypothetical because as soon as a well penetrates the reservoir, such bacteria will have been introduced by means of the drilling mud and completion fluids. As such, it can be safely assumed that such bacteria will be present when a waterflood is conducted because they can be introduced either by means of the drilling or completion fluids or through their presence as spores in the injection water.

SRB are anaerobic bacteria and therefore thrive in the absence of oxygen. Whereas humans use oxygen to assist in the metabolism of food to derive their energy, SRB use sulfate (SO_4^{2-}) as a terminal electron acceptor, reducing it to H_2S.

The food for these bacteria is commonly organic molecules such as small volatile fatty acids (VFAs) with the general formula RCOOH and containing one to four carbon atoms—although the smallest such molecule containing one carbon atom and where R is a hydrogen atom (formic acid) is not typically found in natural systems. SRB typically oxidize these materials to acetate or completely to carbon dioxide (CO_2). VFA molecules can be formed by the anaerobic metabolism of the hydrocarbons in crude oil, and they are commonly found in oilfield formation waters. They are therefore seen as the most commonly used source of food for SRB. However, in some cases, reservoir souring has been encountered in fields where only low concentrations of VFAs are found, such that the level of reservoir souring encountered cannot be explained by VFA metabolism alone (Burger et al. 2013; Zhu et al. 2016; Maxwell and Spark 2005).

It is therefore thought that small aromatic molecules such as benzene, toluene, and xylene (BTX) are also a likely food source for SRB. Further evidence for the role of BTX in reservoir souring comes from the observation that souring occurs despite the limited mixing of seawater (containing sulfate) with formation water (containing VFAs) in most waterfloods and that H_2S continues to be generated over extended periods in oil-saturated sandpack experiments where the only ongoing addition is the injected seawater (Maxwell and Spark 2005).

This suggests that a mixing of the waters might not be needed to induce souring as long as the injected seawater mixes with the oil. The BTX components are preferentially soluble in oil, but they do have some solubility in water. It is expected that if SRB start to consume BTX from the water phase, a partitioning process will be initiated from the oil into the water, replenishing the concentrations in the water phase and thereby providing a sustainable food source for the SRB.

BTX might even be the primary food source for SRB in shallower reservoirs. This is because there is often a relationship between VFA levels in formation water and the reservoir depth, with deeper reservoirs having higher VFA levels than shallow reservoirs (Zhu et al. 2016) (**Fig. 2**). The absence of VFAs at shallow depths can be expected because any VFA originally present in the water will be expected to have been consumed over geological time by the in-situ microbes that are usually found at shallow depths. Thus, it could be that at shallow depths, as BTX is consumed by bacteria, the supply of BTX is continually replenished through the ongoing partitioning from the oil phase, a view that has been supported by a modeling exercise (Burger et al. 2013).

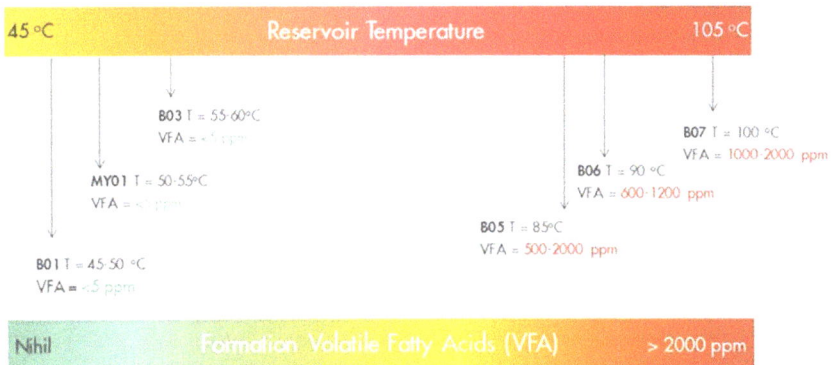

Fig. 2—Observed trends in VFA levels as a function of reservoir temperature (Zhu et al. 2016).

The view that BTX is a food source for reservoir-souring mechanisms is supported by the observed reservoir souring in the Rajasthan Field in India (Burger et al. 2019). In this field, the main water source is produced water, but there is an additional makeup water source with a sulfate content of approximately 500 mg/L. Because the VFA content of the formation water was approximately 20 mg/L, it was not anticipated that reservoir souring would be a problem. Despite this, after less than 4 years of water injection, the H_2S production rate from the field was more than 1000 kg/d and the H_2S concentration in the composite separator gas was approximately 200 ppmv. Modeling was able to show that the observed H_2S production was not possible, even with the complete consumption of the VFAs by SRB. It was then shown that a match of the observed souring levels was only possible when it was assumed that the majority of the inorganic nutrients was being provided by BTX components.

It is also possible that topside processes that use methanol (CH_3OH), such as compact deoxygenation processes, can also serve as a source for SRB metabolism. The commonly recognized food sources for SRB activity are illustrated in **Fig. 3.**

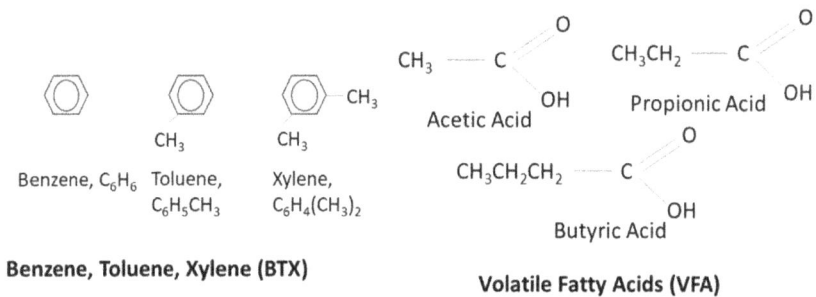

Benzene, C_6H_6 Toluene, $C_6H_5CH_3$ Xylene, $C_6H_4(CH_3)_2$

Acetic Acid Propionic Acid Butyric Acid

Benzene, Toluene, Xylene (BTX)

Volatile Fatty Acids (VFA)

Fig. 3—SRB food sources.

SRB activity and reproduction also require trace concentrations of elements such as nitrogen (N) and phosphorus (P) because they are structural components of some key proteins and nucleic acids, but the required levels of these minerals are low enough that they do not typically present any constraints to SRB activity.

The only other factor needed for SRB to generate H_2S is sulfate (although it should be noted that SRB might be able to use other sulfur-containing compounds to support their metabolism). Sulfate can be present in some formation waters. Usually, however, the source of the sulfate seen in reservoir souring is the injection water. Most commonly, sulfate is found in seawater floods. The concentrations of sulfate in seawater vary somewhat depending on the geographical location but are typically observed in the range of 2400 to 3200 mg/L—a more than adequate concentration to support SRB activity.

Because the presence of SRB is generally a given, reservoir souring should be expected for all cases where sulfate is present and an appreciable food source from either VFAs or small aromatic compounds is present. This is now almost universally recognized, although it has taken some time to reach this point. It was not until a large number of waterfloods suffered reservoir-souring problems in the 1980s that these factors began to be widely recognized.

This is despite the fact that seawater injection was first used for waterflood purposes in California in the 1950s. A seawater-injection scheme began in the Wilmington Field in 1956 (Fernandes 1956) after a number of other industrial users in California—including power plants and refineries—began using seawater. Today, it might come as no surprise to hear that this field appears to have soured (Miller and Robuck 1972), although the link between the occurrence of reservoir souring and the presence of sulfate in the injection water was clearly not recognized in the early waterfloods. It is not known if this was the first waterflood to have suffered from souring because there could have been earlier projects that soured from the use of other sulfate-containing injection sources.

It is important to recognize that the generation of H_2S by bacteria in the reservoir does not automatically imply that H_2S will be produced. This is because the generated H_2S needs to be transported through the reservoir to the producing wells. That might be likely to occur because the waterflood is a displacement process that aims to drive fluids toward the producing wells, but it is not a given. These factors will be discussed in additional detail in the coming sections.

3.3 Reservoir-Souring Mechanisms. Two primary mechanisms have been proposed to explain how SRB generate H_2S in waterflooded fields (**Fig. 4**). The first attempt to describe a reservoir-souring mechanism came in 1991 with the publication of the mixing-zone souring model (Ligthelm et al. 1991) in which it is assumed that bacterial activity takes place deep within the reservoir. In this model, the sulfate is provided by the injected water, and the carbon source (VFA) is provided by the formation (connate) water. Consequently, it is expected that H_2S will be generated at the flood front. Because H_2S is generated in the water phase, it can be expected that it will be transported toward the producers in line with the movement of the injection water. During that transportation, the H_2S might interact with the oil and the iron-containing minerals within the reservoir rock. This interaction might reduce the development of H_2S concentrations at the producing wells to some extent, but generally speaking, H_2S production will be expected at, or very soon after, injection-water breakthrough.

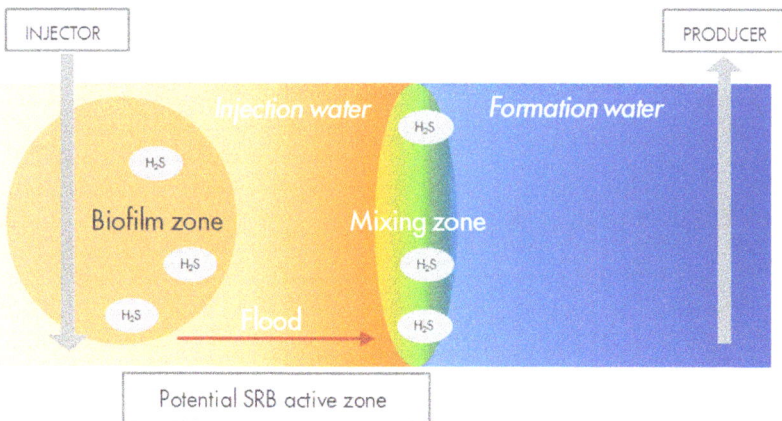

Fig. 4—Reservoir-souring mechanisms (Zhu et al. 2016).

It is evident that this model requires that the SRB are able to thrive in the deep reservoir conditions. One reservoir parameter that could constrain that ability is temperature. There is, unfortunately, uncertainty as to what reservoir temperature might act as a limitation to SRB activity because there are a number of different SRB species that can cause souring; each could have different temperature tolerances, and it might not be clear which species are present in any individual application. Generally, because thermophilic bacteria are typically found to have upper temperature tolerances of approximately 70 to 80°C, such temperatures are therefore likely to represent the upper limit for the development of mixing-zone souring.

It is also possible that high formation-water salinity could act as a constraint, at least temporarily, because most SRB species are inhibited by high salinity. Tests conducted for an application in Kuwait indicated that SRB growth was inhibited at total dissolved solids (TDS) levels above approximately 120 000 mg/L (Al-Rasheedi et al. 1999). Even so, there remains a great deal of uncertainty as to the level of salinity that will act as a constraint to SRB activity. This is because there is such a large range of SRB species that might be active within a reservoir. SRB species originating from fresh water might be inhibited by salinities of approximately 20 000 to 30 000 mg/L of sodium chloride (NaCl), but, by contrast, many marine SRB species are considered mildly halophilic and might require 10 000 to 30 000 mg/L of NaCl to achieve optimal growth. The activity of most SRB will decline significantly if the NaCl concentration exceeds 50 000 to 100 000 mg/L (Postgate 1984).

However, as the flood proceeds, it might be expected that, even if the original formation-water salinity is high, over time the water salinity in the pore space will drop because the original connate water will gradually be replaced by injection water, especially for cases where significant numbers of pore volumes are injected. As such, salinity might only act as a constraint to mixing-zone souring early in the life of the waterflood.

In a mixing-zone scenario, the backflowing of injection wells would not be expected to result in observable levels of H_2S because even if SRB are present at that location, the very limited food source available in seawater would not lead to any appreciable H_2S generation. On the other hand, it might be expected that after formation-water breakthrough occurs, there will be a depletion in the original VFA levels in formation water. This was observed during the souring of the Alba Field in the North Sea (Dunsmore and Evans 2006).

The fact that reservoir souring has occurred in fields where reservoir temperatures suggest that mixing-zone souring would not be possible indicates the need for an alternative souring model for such cases. Furthermore, H_2S production often lags behind injection-water breakthrough, sometimes very significantly, and this also suggests some limitations associated with the mixing-zone model. This subsequently led to the development of the biofilm souring model (Sunde et al. 1993), which assumes that SRB activity is limited to the relatively cooler areas around the injector wellbore(s). For this model, the rate and extent of reservoir souring will be governed by

- The availability of nutrients: These nutrients can come from the injection water itself, and the VFAs in the near-wellbore region might be consumed. However, these sources could be limited, which might suggest that residual oil is being used as a food source. In this case, BTX might be a very important food source because, as it is consumed, additional replenishment will be expected to occur through partitioning from the oil into the water, thus providing a sustainable food source.

- H_2S absorbance: Any capacity that the reservoir rock has to absorb H_2S, which will increase the number of pore volumes injected before the H_2S will be observed in producers.

It is evident that the onset of H_2S production will lag behind the injection-water breakthrough for the biofilm souring model. One key difficulty for this model is that seawater contains only limited amounts of nutrients. However, it is possible that, for some cases at least, poor control in surface injection facilities could introduce additional nutrients. For example, methanol is used in compact catalytic deoxygenation processes, and leakage of this chemical would undoubtedly provide additional nutrients for SRB.

For fields undergoing seawater injection where the onset of H_2S production is significantly retarded compared to seawater breakthrough, this is very strongly suggestive of biofilm souring. The Gullfaks Field, in the Norwegian North Sea, is an example of such a case (Mitchell et al. 2017). In this field, H_2S production was observed shortly after water breakthrough, suggestive of mixing-zone souring, but it was followed by a subsequent decline in H_2S levels. Several years later, a gradual increase to much higher levels was recorded, and the implication is that this latter H_2S production was associated with biofilm souring (**Fig. 5**).

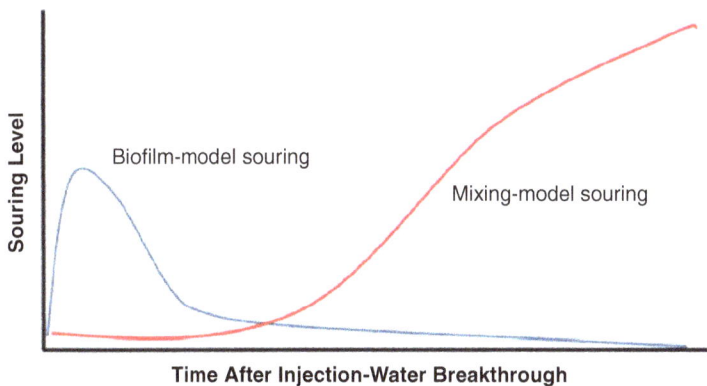

Fig. 5—H_2S trends where both biofilm and mixing-zone souring occur (Mitchell et al. 2017).

In a field where biofilm souring is the dominant souring mechanism, the backflowing of injection wells can be expected to result in an initial water flow that is likely to contain H_2S and be heavily infested by bacterial slime, and might also appear black as a result of the presence of iron sulfide. There are cases that do confirm the presence of near-wellbore biofilms (Cusack et al. 1987), which suggests that when the near-wellbore formation is plugged by bacterial biofilm with large amounts of entrained inorganic particulate material, the biofilm must be removed with bleach before any damage attributable to particulates can be removed by an acidization treatment.

A weakness of the biofilm model is that it does not explain how it is possible to get significant levels of reservoir souring in cases where seawater is injected below the oil/water contact and where limited amounts of organic compounds available for SRB growth are expected.

A further adaptation of the main souring models is the thermal viability shell (TVS) model (Eden et al. 1993), which is based on the adaptation of mathematical models built to predict the temperature distribution around injection wells. This model predicts that the likelihood of souring in a reservoir is dependent on the establishment of a stable viability shell in either the mesophilic (20 to 40°C) or thermophilic (40 to 80°C) SRB temperature ranges. It suggests that souring from thermophilic sources is potentially more serious than that from mesophilic sources in North Sea reservoirs. Indeed, because the vast majority of reservoirs have temperatures greater than 40°C and because the cooled zone around injection wells might usually be expected to be small in comparison to interwell distances, this might be expected to be true for the majority of waterflooded fields.

The TVS model does not consider the impacts of nutrients on H_2S generation, nor does it take into account any effects resulting from adsorption or partitioning. Nevertheless, it is useful in considering where in the reservoir H_2S is being generated, as well as the impact this might have on when H_2S is likely to be produced.

As the data from Gullfaks suggested, it is likely that both mixing-zone and biofilm souring models are at work in many field cases. Reservoir souring of up to 1,000 ppmv in the gas phase was experienced in producers in the Fulmar Field in Scotland despite the fact that the reservoir temperature, 127°C, is almost certainly beyond the temperature tolerance of all SRB.

These souring models suggest that all seawater floods, or indeed any waterflood in which sulfate is present in the injection water, will suffer from reservoir souring. The only uncertainty appears to relate to whether the generated H_2S will be produced and at what concentration. It has been suggested that the biofilm model, rather than the mixing-zone model, more accurately represents the conditions that cause reservoir souring but that further refinement of the assumptions might be required to make a more complete prediction of souring behavior (Maxwell and Spark 2005).

3.4 Factors Impacting Souring Severity and Timing. The way that SRB metabolize VFAs in the presence of sulfate suggests that reservoir souring is almost certain to occur in seawater floods. The only question that remains is when the H_2S will appear at the producing wells and how bad the problem will become. Part of the answer to this question comes from an assessment of whether it is biofilm souring or mixing-zone souring that is likely to be the dominant souring mechanism. The reservoir temperature is the most critical factor, and it is likely that H_2S production will occur earlier when temperatures are low enough to facilitate mixing-zone souring (assume temperatures below 70 to 80°C).

There are, however, a number of other complicating factors that will further influence the answer to this question. One such factor is the injector location. If the mechanism is biofilm souring, injectors that are placed down in the water leg might experience an even-further-delayed onset of H_2S production because much of the generated H_2S might never escape from the aquifer leg.

Of course, it is not a question of whether H_2S is being generated; the issue is whether it is transported through to the producers. This suggests that H_2S is more likely to occur in reservoirs where the rock is more oil-wet, or in cases where the mobility ratio is unfavorable, because these scenarios are more likely to experience higher pore-volume water throughputs, and those multiple pore volumes of injection are expected to result in an increased likelihood of the generated H_2S being transported to the producers.

The partitioning of H_2S also plays a role. Because the H_2S in waterfloods is generated by SRB activity, it is generated in the water phase, but it will subsequently begin to partition between the oil, water, and (if present) gas phases. Thus, in cases where a gas cap is present, for example, some of the generated H_2S could end up in the gas cap. Some generated H_2S will also end up in the residual oil (Maxwell and Spark 2005).

Generally speaking, H_2S is considerably more soluble in organic compounds than it is in fresh water or brines. This solubility increases with an increase in aromatic nature and decreases with an increase in paraffin and the polar nature of the hydrocarbons. In previous years, the quantification of partitioning coefficients was complex because H_2S gas often exhibits nonideal gas behavior, demanding the use of cubic equations of state. However, these are readily incorporated into current models, making this a much easier exercise for the modern engineer.

One important fact to note here is that as the producing water cut rises, the volume of produced gas will start to appreciably decline. This means that the H_2S will be partitioning into a smaller and smaller gas phase—and it is in this phase that the H_2S measurements are most commonly made. Consequently, at high water cuts, the measured H_2S concentrations in the gas phase will start to rise. This will occur even if there is no overall increase in the amount of H_2S produced, and the phenomenon occurs purely as a result of the decrease in produced gas volume. This fact must be considered when interpreting produced H_2S data. An example of this effect is shown in **Fig. 6.**

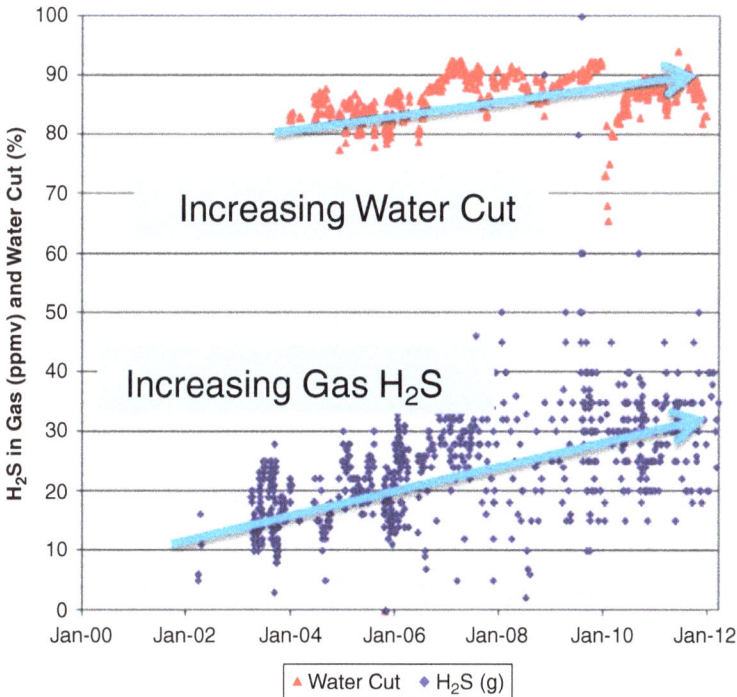

Fig. 6—H_2S in gas increase as a function of increasing water cut (after Evans et al. 2015).

Furthermore, the presence of iron-bearing minerals in the reservoir rock has the potential to scavenge any H_2S that is generated in situ, thereby limiting the extent of the problems encountered in the producing wells. The mineral siderite ($FeCO_3$, iron carbonate) is known to be particularly effective in this respect (Burger and Jenneman 2009), but other iron-containing minerals might also exhibit some beneficial scavenging effects. The reactions of siderite, hematite (Fe_2O_3), and magnetite (Fe_3O_4) with H_2S, showing how they help to limit the problems, are shown:

$$FeCO_3 + H_2S \rightarrow FeS + H_2O + CO_2. \dots\dots\dots\dots\dots\dots\dots\dots\dots\dots\dots\dots (8)$$

$$Fe_2O_3 + 3H_2S \rightarrow FeS + FeS_2 + 3H_2O. \dots\dots\dots\dots\dots\dots\dots\dots\dots\dots\dots (9)$$

$$Fe_3O_4 + 4H_2S \rightarrow 2FeS + FeS_2 + 4H_2O. \dots\dots\dots\dots\dots\dots\dots\dots\dots\dots\dots (10)$$

Some reservoirs contain appreciable amounts of these minerals, and siderite is an especially common reservoir mineral. Ideally, the distribution of the minerals throughout the reservoir will be known because a mineral is likely to be more effective at scavenging if it is located throughout the reservoir. Similarly, its form in the reservoir could be important—nodules might be less effective scavenging agents than a mineral that is coated on sand grains and thus has a large associated surface area. A waterflood in the Champion Field in Brunei operated for more than 30 years before H_2S was observed in any producing wells because the generated H_2S was mitigated by the presence of extensive (approximately 5%) amounts of siderite.

Another extremely important factor that influences the severity of reservoir souring is the decision regarding whether the produced water will be reinjected. Seawater injection provides a significant source of sulfate ions, and if produced water is also injected, a significant source of the molecules used for metabolism by SRBs is often provided. This creates a perfect combination for severe reservoir souring; mixed seawater/produced-water injection systems often exhibit the most severe levels of souring and will inevitably sour more severely than the seawater-only scenario. To illustrate the impacts, the predicted levels of reservoir souring in the Ekofisk Field in the Norwegian North Sea under seawater injection have been compared to the case in which seawater is injected with produced water. Burger et al. (2006) predicted that H_2S in separator gas would rise to a maximum of approximately 28 ppm under seawater-only injection but to approximately 45 ppm under seawater injection with produced-water reinjection (PWRI) (**Fig. 7**).

Burger et al. (2006) also illustrate how only a portion of the generated H_2S gets produced from the reservoir. In this case, only approximately 10% of the H_2S generated gets produced (**Fig. 8**), with the remainder left in the formation water and residual oil.

It is important to recognize that the increased souring risk for PWRI systems only applies in cases in which the produced water is injected with another water source that contains appreciable sulfate concentrations. In cases in which only produced water is injected, the SRB activity is likely to be limited by the limited amount of sulfate present. However, even where the produced water contains no sulfate, some reservoir souring might be encountered because the topside process can reintroduce sulfate from other sources such as tetrakis(hydroxymethyl)phosphonium sulfate (THPS) biocide or a chemical oxygen scavenger, for example.

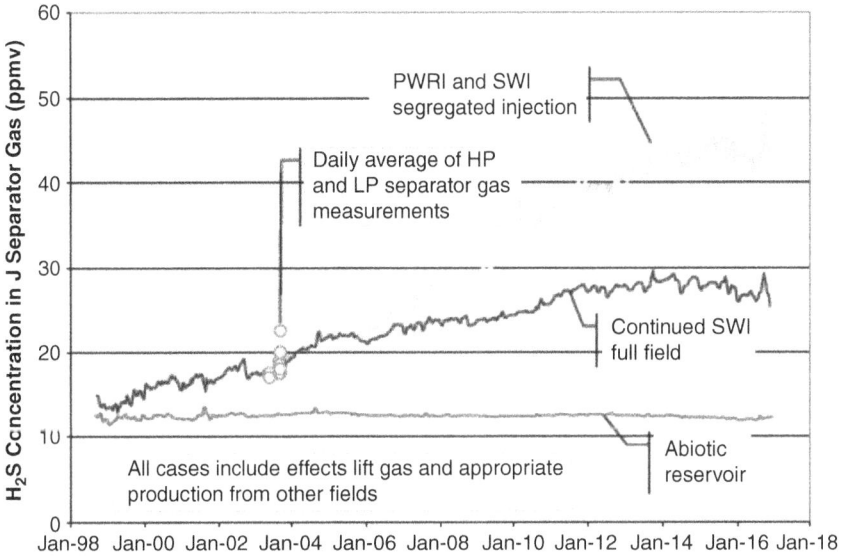

Fig. 7—Forecast separator-gas H_2S content at Ekofisk (after Burger et al. 2006). SWI = seawater injection, HP = high pressure, and LP = low pressure.

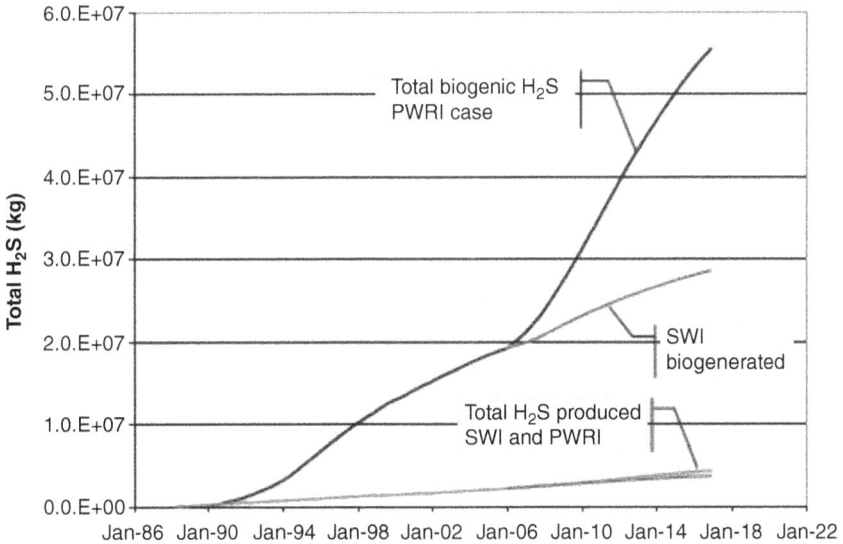

Fig. 8—Cumulative H_2S generation and production at Ekofisk (after Burger et al. 2006). SWI = seawater injection.

There is one last factor to consider that, in some circumstances, might work to limit the extent of reservoir souring. This is the fact that the H_2S generated as a waste product by SRB activity is actually toxic to those bacteria. The inhibiting effect of sulfide is probably dependent on the species of SRB present and might also be a function of the pH of the water. One study reported a 50% inhibition of the activity of mesophilic SRB at 240 ppm total sulfide and 83 ppm undissociated H_2S, while Reis et al. (1992) reported that in a pH range of 6.2 to 6.7, H_2S completely inhibited the culture growth at an H_2S concentration of 547 mg/L. McCartney and Oleszkiewicz (1991) found that both propionate removal and sulfate reduction were inhibited in direct proportion to the concentration of sulfide.

Such results clearly demonstrate the toxicity of H_2S to SRB. It must be recognized, however, that the H_2S-tolerance limits of bacteria as reported in the literature might not be high enough to prevent the activity of SRB in the parts of the reservoir otherwise conducive to their growth because most of the bacteria could still be active as a result of the potentially protective properties afforded by biofilms.

The complexity of such constraints illustrates the extreme difficulty associated with quantifying how severely a reservoir might sour under waterflood. What is evident is that increases in sulfate levels and increases in carbon sources such as VFAs and BTX are likely to result in increased levels of souring and that measured levels of H_2S in gas are likely to increase late in field life as the generated H_2S partitions into an ever-decreasing gas phase.

In summary, even though H_2S generation is expected for most seawater floods, H_2S production can be less likely in cases with injection into the water leg of the reservoir and where a piston-like flood displacement occurs such that the injection of only a small number of water pore volumes is required. By comparison, the injection of water into the oil leg of the reservoir and the need for the injection of a large number of pore volumes of water—such as in cases where the microscopic displacement is poor (e.g., in viscous oil reservoirs)—are more likely to result in appreciable H_2S production.

3.5 Souring Modeling and Prediction. The previous sections might suggest that reservoir-souring modeling and prediction is an art rather than a science because of the number of factors that limit the ability to accurately predict souring levels. These factors include

- The number of souring models: There are two (or three?) souring models, and there is an uncertainty regarding which one to apply (or are both types taking place?). In cases where biofilm souring is believed to be at work, what is the depth away from the injection well that the biofilm can penetrate? This will depend on the species that are present and their temperature tolerance with respect to the injection-water and reservoir temperatures.
- An incomplete understanding of the reservoir and its flow paths (when will breakthrough occur?): Most modeling exercises use the reservoir model as a base. Because water breakthrough occurs before the model prediction, more often than not, there is already some uncertainty associated with souring timing and severity. To account for this problem, modeling should ideally be performed using a range of static models representing the range of potential geological realizations, thus producing a range of potential souring outcomes.
- An incomplete understanding of the amount of food available to SRB.
- An incomplete knowledge regarding which SRB species are active within the reservoir and where they are active.

- Limited knowledge regarding the reservoir scavenging capacity.
- Uncertainty associated with H_2S partitioning between oil, water, and gas phases.

These factors contribute to appreciable uncertainty in the modeling process and suggest that H_2S forecasts should be viewed only as a guide to the expected souring severity.

On the basis of this information, it might be appropriate to first look at analogs when considering how badly a field might sour, although careful thought is needed to define exactly what constitutes an analog.

Where a modeling exercise is being considered, it is first appropriate to consider if it is possible to model key factors such as biogenic activity, H_2S partitioning, and H_2S scavenging using the same simulator as that used to predict oil, water, and gas production. Because the answer to this question is usually no, it is necessary to link the reservoir simulator to a reservoir-souring simulator. Even then, because there is a significant amount of uncertainty associated with the expected extent of reservoir souring, it is also appropriate to conduct sensitivity analyses, thereby generating a range for the extent of expected H_2S values.

A number of different models are available in the industry. Such models have typically been history matched to previous field cases. These models are able to predict H_2S on an individual-well basis for the oil, water, and gas phases, and they might also feature a mitigation module for nitrate injection. Of course, the predictive capability of such models is not only a function of the accuracy of the souring prediction module but is also dependent on the accuracy and representativeness of the reservoir model itself.

One 3D reservoir-souring simulator incorporates the effects of key variables such as temperature and nutrient supply, as well as the interaction of the generated H_2S with reservoir rock and the partitioning of the H_2S between the relevant phases (Delshad et al. 2009). This model has subsequently undergone additional refinement to incorporate temperature, salinity, and pH and their overall effects on biological activity (Hosseininoosheri et al. 2017).

Possibly the most widely used model is one that was developed in a joint industry project (Dunsmore and Evans 2006). It requires a full-field reservoir simulation model with history-matched oil, water, and gas production rates per well and is compatible with a number of standard reservoir simulators. This model was developed as a "post-process" to existing reservoir simulators in that it takes the output of a simulation run and processes the results using its own souring model kernel. Therefore, an assumption in this model is that biochemical reactions do not influence mass transfer. It is able to

- Model mixing-zone and biofilm souring (and is able to specify the depth to which the biofilm can penetrate)
- Take into account the impacts of all potential food sources on the extent of H_2S generation (including the assimilation of components from crude oil)
- Capture the impacts of nitrate injection
- Take into account the H_2S scavenging capacity of reservoir rock
- Take into account the mass-based H_2S partitioning coefficients for gas/oil and oil/water as a function of pressure, temperature, pH, water salinity, and H_2S partitioning

This model was used in an ongoing prediction of reservoir souring in a mixed produced-water/aquifer-water injection scheme in the North Sea where, after achieving a good history match based on production up to 2009, the ongoing H_2S levels were seen to be consistent with the model predictions up to 2012 (Evans et al. 2015). Al-Refai et al. (2019) reported that a simulation using this model has been used for a Kuwaiti field.

The model can also simulate laboratory-sized bioreactor columns, so it is also possible to provide a simulation of field-scale H_2S levels based on laboratory-scale experimental data.

The modeling of reservoir souring is beginning to improve but, because of the high level of complexity and the large number of moving parts associated with the souring process, such models still carry a fair degree of uncertainty.

Given what has been said thus far about reservoir souring, the first step in assessing the reservoir-souring risk for a new development is to create a simple risk assessment based on the key risk factors. These include

- The location where the H_2S is being generated—is it only near the injection wells, or is it likely to be generated throughout the reservoir?
- The number of pore volumes of water to be injected because the larger this number is, the more likely it is that H_2S will be transported to the producers. Larger pore-volume throughputs are more likely for heavy-oil developments, heterogeneous reservoirs, and cases where the reservoir rock is primarily oil-wet.
- The conditions of temperature and salinity, and how they impact biological activity.
- The amount of food available to support biological activity.
- The location where the water is being injected—is it peripheral, or is it into the oil column?
- The amount of material present in the reservoir that might act as a natural scavenger for any H_2S that is generated.
- The amount of water, gas, and residual oil present in the reservoir for H_2S to partition into.

For new field developments, it might be appropriate to perform a souring modeling exercise if the initial assessment has suggested a medium to high souring risk. It should be remembered that any analysis will carry a reasonably high degree of uncertainty and that the model should be calibrated and updated after the project moves into the operating phase.

3.6 Souring Prevention. Because it is agreed that reservoir souring is occurring by means of a biological process driven by the activity of SRB, it is therefore evident that any attempt to control reservoir souring needs to be based on methodologies designed to control the activity of those bacteria (Jones et al. 2018). One available philosophy that is widely used is to effectively limit all biological activity through the addition of chemicals that inhibit that activity. An alternative strategy is to artificially change the natural balance of chemicals present so that the activity of SRB, in particular, will be negatively impacted. Each of the relevant strategies is discussed in more detail in the sections that follow.

3.6.1 Biocide Treatment. Biocides are invariably applied in water-injection systems to limit biological activity. Most often, the prime purpose of biocide-injection schemes is to control bacterial activity within the topside water-injection process. This is required because

- The cells of bacteria have mass and therefore contribute to the solids loading of injection water; this will negatively impact the injectivity of the water-injection wells.
- If biofilms are allowed to develop, they will facilitate the initiation of under-deposit corrosion. The high flow rates observed in water-injection systems will also facilitate the periodic sloughing-off of parts of the biofilm, which will be expected to have a negative impact on well injectivity.

It is also appropriate to consider which biocide to use for bacterial control. Historically, a wide range of biocides have been used for bacterial control in seawater-injection schemes. Nearly all facilities use chlorination to disinfect the seawater intake system, but because of the toxicity of chlorine gas, the chemical is invariably dosed as a solution of sodium hypochlorite, the active constituent in household bleach. This is typically generated on-site in a simple electrolysis cell. However, chlorine reacts with the chemical oxygen scavenger that is normally used to deliver the oxygen specifications needed to avoid corrosion in the water-injection system. As a result, there is no residual biological control downstream of the injection point for that chemical, and therefore, the hypochlorite-injection program will have no impact on the prevention of reservoir souring.

Because of the interaction between hypochlorite and the chemical oxygen scavenger, a supplemental biocide-injection program is applied to nearly all water-injection systems to ensure topside biological control downstream of the oxygen-removal process. This supplemental biocide injection might have some impact in limiting reservoir souring, but it is not certain. A wide range of chemicals, discussed in the sections that follow, has been used for this purpose.

Aldehydes. The most commonly used aldehydes are formaldehyde and glutaraldehyde. Aldehydes are effective general biocides, but their penetration into biofilms is poor as a result of inactivation by the bacterially generated polysaccharide cholylglycine. Because of this, aldehydes are sometimes blended with quaternary ammonium compounds or amines to assist in penetration. However, because of the toxicity of these products and the fact that they are believed to be carcinogenic, they are no longer widely used.

Amines. Alkyl-substituted amines are surface active, forming films on exposed metal and having general corrosion inhibition properties. They have also been blended with aldehydes to make best use of the different properties of the two types of chemicals.

Isothiazolones. Isothiazolones are not usually the first biocide choice for water-injection systems, although work has been published showing them to be effective when used as frequent shock doses or as a continuous treatment.

Thiocyanates. Methylene bis(thiocyanate) has been widely used for biological control in paper mills and has occasionally been used as a water-injection-system biocide. Its stability and activity decrease with an increase in pH. Its solubility in seawater is poor, but it is a low-toxicity biocide, which could be advantageous in environmentally sensitive areas.

Anthraquinone. Anthraquinone is another option for control of SRB. The key feature that differentiates it from other control mechanisms is that it is a bio-stat—a chemical compound that halts one of the respiration pathways without harming the organism. Biostats stop the formation of certain detrimental metabolic products such as acids and H_2S through the halting of the respiratory pathway of the bacterial organism. Anthraquinone uncouples the electron transfer process in SRB that is required for the bacteria's respiration using sulfate (Burger and Odom 1999).

This biocide can be considered for use in cases where nitrate injection is applied because it has been suggested that its use could allow for lower nitrate-injection rates to be applied (see Section 3.6.2). It has also been claimed that it increases the rapidity with which the microbial population in the reservoir changes from sulfate to nitrate reduction.

Acrolein. Acrolein is a highly reactive acrylic aldehyde (propenal). It is a volatile liquid with a pungent odor, and it is very difficult to work with. It is, however, an effective biocide at low doses for short contact times. (Less than 25 ppm for 1 hour has been effective against sessile bacteria.) In addition to having biocidal properties, acrolein is an active sulfide scavenger (Kissel et al. 1985) and has been used for this role. Acrolein can be generated on-site from inactive, nonhazardous precursors using a catalytic column immediately before injection into the system.

Over the years, there has been a change in the type of biocides most commonly used for bacterial control in waterfloods. The reasons for this change include the toxicity and environmental impacts described, but another is that a number of biocides exhibit declining performance over time because bacterial populations often start to develop a tolerance to the applied chemical. (Because of this, assets historically have tended to use two different biocides, alternating their injection.)

The most commonly used biocide in water-injection systems today is almost certainly THPS, which is a quaternary phosphonium salt. This product was first used in cooling towers, where it was demonstrated to be a rapid-acting, high-performance biocide. Consequently, it was first used in water-injection systems in the late 1980s, and it performed so effectively that it has since become the primary biocide used in water-injection systems (Jones et al. 2012). It shows high rates of biodegradation, so the environmental risks associated with accidental discharge are relatively low. Another critical factor behind the popularity of this chemical is that it has more than one mode of action against bacteria, and as a result, there is a significant reduction in the ability of the bacteria to develop a tolerance. There is therefore no need to alternate this chemical with a different biocide, which makes it somewhat more attractive operationally.

Indeed, there are three identified modes of action for THPS, which minimizes the chance that bacterial resistance will be developed:

- It has a direct effect on cell-wall proteins.
- It inhibits lactate dehydrogenase, a key enzyme.
- It inhibits the adenosine diphosphate to adenosine triphosphate conversion, a key respiratory pathway.

The first of these mechanisms provides the primary, high-concentration acute effect, disrupting cell structure and causing lysis, while the other two are longer-term, lower-concentration effects.

It is perhaps self-evident that biocides have the ability to kill bacteria; therefore, it might appear that because reservoir souring in waterfloods is a biological phenomenon, biocides at least have the potential to control it. One key problem, however, is that there is little to no consensus regarding the required biocide dosage rates needed to control bacterial activity.

THPS biocide replaced the previously applied weekly glutaraldehyde batch treatments in Chevron's Ninian Field in the UK North Sea after the field had soured, to try to keep the topside clean (Macleod et al. 1994). Although the THPS product was four times more expensive than the glutaraldehyde previously used, a reduced-dosage regime (100 ppm for 3 hours per week rather than 250 ppm for 6 hours per week) allowed for the change of product while reducing operating expenses. After using the THPS product for 1 year, vastly improved biological control was noted in the topside and H_2S levels in the producers stabilized. Monitoring for the biocide took place in the producers, and peaks were seen on a weekly basis, indicating that the biocide was traveling all the way through the reservoir (although the sharpness of the peaks suggested that the flow pathway through the reservoir was occurring by means of a short circuit).

Oduola et al. (2009) reported that twice-weekly THPS doses of approximately 450 ppm are needed to mitigate SRB. However, different bacterial populations will be present in different fields, and they can be expected to have different tolerances to biocides. Furthermore, different process chemicals are used in different water-injection systems, and those different chemicals might be expected to have different levels of interference with the biocide that is used. Consequently, although a biocide dosage program might be optimized at the topside on the basis of the outcomes of surveillance of the biological populations, the difficulty in determining the required biocide program for topside control suggests that it could be far more difficult to fully quantify the required dosage regime to protect all of the reservoir.

When THPS is used as the biocide, it might be extremely difficult to effectively control souring for fields where some degree of souring has already been encountered. This is because the THPS will react with H_2S; while there might be some beneficial scavenging effects, it could be almost impossible to determine the required dosage for souring control.

There might then be a more pragmatic limitation in the ability of biocides to control reservoir souring. Although all waterfloods invariably use biocides, it is not uncommon to see processes that are severely infected by bacterial populations. The primary reason for this is almost certainly cost. A biocide program is extremely costly; indeed, this is one of the reasons that this chemical is usually injected on a batch basis rather than injected continually. There is often pressure from management to limit the cost of such programs, and this can have the effect of limiting their effectiveness. It should be evident that if a field is unable to control bacterial activity within the limited confines of the topside injection facilities, the same philosophy will not be effective in controlling bacterial activity throughout the entire reservoir.

Thus, although biocide control might be a plausible control methodology in theory, it is a difficult means to effect control in practice. Having said that, because it is now well-established that SRB activity is responsible for reservoir souring, it is logical to assume that robust biological control can be expected to at least limit bacterial activity in the reservoir and therefore will have some sort of controlling impact on the degree of reservoir souring.

There is some evidence that biocide treatment can have a positive effect on the extent of reservoir souring. Larsen et al. (2000) reported that batch treatment of THPS at dosages between 200 and 400 ppm was effective in reducing H_2S in the short-term in Denmark's Skjold Field. These trials suggested that, to be effective, THPS needed to be applied as a formulated package with a surfactant biopenetrant because one trial was conducted without the added surfactant and did not have a positive effect on downhole H_2S levels.

The positive effects that THPS had could easily be seen by means of the (negative) impacts on producing H_2S concentrations when the field reverted to the previously used biocide (**Fig. 9**). However, Skjold is a heavily fractured reservoir, and it appears likely that the THPS was short circuiting most of the reservoir through this fracture network. It remains unclear if such results would be replicated in cases where all the water has to pass through the reservoir matrix.

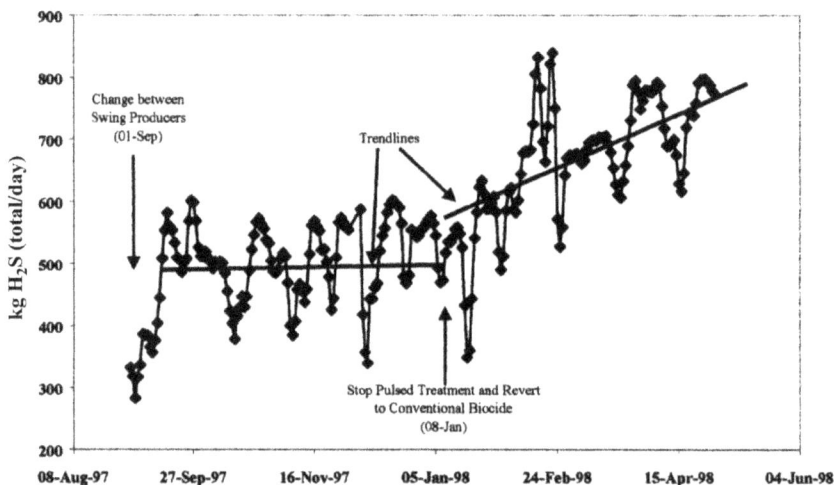

Fig. 9—Effect of changing from a THPS pulsed treatment to a conventional biocide (Larsen et al. 2000).

Some operators tend to rely solely on biocide treatment for biological control, although the value of supplementary souring control is becoming increasingly recognized. Where biocide-only control for reservoir souring is used, a robust control program is needed. The approach typically entails a continuous biocide injection at the outset of the project (20 to 25 ppm might be suggested for the first 3 months) to prevent initial biofilm development. Ongoing control is then achieved through continuous chlorination, although this might need to be delivered at the inlet to the de-aerator as well as at the normal seawater-intake injection point, and would be supplemented with ongoing batch treatments of proprietary biocide. The requirements for such treatments will almost certainly be case specific in terms of the type of biocide used as well as the applied dosage, duration, and frequency. An initial dosage of 250 to 500 ppm for 4 hours twice a week could be an appropriate starting point, but an ongoing bacteria monitoring program is a critical requirement so that the treatment regime can be optimized on the basis of the results of that surveillance. Particular attention might need to be paid to the filtration units, and dedicated biocide treatments might need to be applied at the end of backwash periods on the filters.

It might be argued that this procedure simply represents sound operating practice. However, the historical difficulty many operators have experienced in paying adequate attention to waterflood operations, combined with the severe consequences of failing to maintain adequate souring control, suggests that a reliance on biocide control alone to prevent reservoir souring might be inadequate. It is also worth noting that many fields have suffered from reservoir souring and nearly all have used some degree of topside biological control. Furthermore, relying on biocide alone can be an expensive option.

The view of many operators is that although biocides can limit biological activity, they cannot fully control it. That statement would appear to be true for topside systems and is likely to be an even more accurate statement when applied to downhole bacterial control. There is also uncertainty regarding the ability of biocides to exert control throughout the reservoir (except, as has already been noted, where there are reservoir short circuits). As a result of these issues, operators that tend to use biocide-only protection are beginning to consider the need for additional protection in high-risk applications.

Indeed, there is a field case from the North Sea where a degree of reservoir souring has been observed despite continuous injection of THPS biocide (Evans et al. 2015). This field used a robust, continuous biocide dosing program with 40 ppm (based on well water volumes) injected into the production header and an additional 40 ppm injected downstream of the produced-water booster pumps. These dosages suggest that by the time the water reaches the injection pumps, it has been continuously dosed with 80 ppm of THPS.

There are a number of key features associated with this case. First, the reservoir temperature in this field is only 40°C, which indicates that mixing-zone souring is possible. Additionally, injection used a mixture of produced water and aquifer water that had a low sulfate content—typically of approximately 10 mg/L. Because the organic content of the produced water is in the range of 30 to 42 mg/L, which is sufficient to support significant biological activity, it is clearly the sulfate content that limits the biological activity.

H_2S concentrations in the gas phase measured at the free-water knockout tanks have ranged from 10 to 50 ppmv. This concentration is linked to the water cut, and the underlying H_2S concentration in the water phase appears to be steady at 2 mg/L, suggesting a low but consistent level of reservoir souring. Simulation studies suggested that the continuous biocide dosing had little impact on the observed levels of souring, and therefore, continuous biocide dosing was ceased and a batch biocide program initiated. This case history highlights the limitations of continuous biocide dosing as a reservoir-souring control strategy, especially for reservoirs prone to mixing-zone souring.

3.6.2 Nitrate Injection. Nitrate has long been used to control the unpleasant smell associated with low concentrations of H_2S in sewage systems, which led to it first being considered in the 1980s as a potential means to control reservoir souring in oilfield systems (Jack et al. 1984; Jenneman et al. 1986; Hitzman 1994). In recent years, nitrate injection has emerged for many companies as the standard method to control reservoir souring, at least for seawater floods. This chemical, commonly injected as either the calcium or sodium salt, has the potential to inhibit reservoir souring by a number of different mechanisms:

- Through biocompetition, nitrate injection enhances the activity of nitrate-reducing bacteria (NRB) at the expense of SRB. These bacteria use the same food sources as SRB, and by promoting the growth of the NRB, it has been suggested that the carbon sources are then unavailable for the SRB. This mechanism is possible because nitrate/nitrite and nitrate/nitrogen redox couples are thermodynamically much more favorable than sulfate/H_2S:

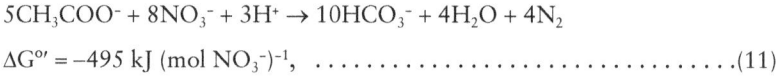

$$5CH_3COO^- + 8NO_3^- + 3H^+ \rightarrow 10HCO_3^- + 4H_2O + 4N_2$$

$$\Delta G^{o\prime} = -495 \text{ kJ (mol } NO_3^-)^{-1}, \dots\dots\dots\dots\dots\dots\dots\dots\dots\dots\dots (11)$$

but

$$CH_3COO^- + SO_4^{2-} + 3H^+ \rightarrow 2HCO_3^- + HS^-$$

$$\Delta G^{o\prime} = -47 \text{ kJ (mol } SO_4^{2-})^{-1}, \dots\dots\dots\dots\dots\dots\dots\dots\dots\dots\dots (12)$$

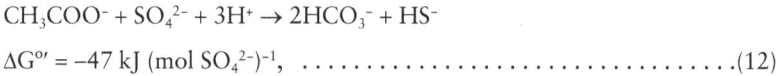

where $\Delta G^{o\prime}$ is the Gibbs free energy (also sometimes called the free enthalpy), which is the maximum amount of nonexpansion work that can be extracted from a thermodynamically closed system in the standard state.
- It has been suggested that many oil and gas fields contain NRB and sulfide-oxidizing nitrate-reducing bacteria (SONRB) populations and that these are activated upon the injection of nitrate or nitrite, allowing sulfide removal in situ (Jenneman et al. 1997). Thus, SRB use the reducing equivalents in the organics to reduce sulfate to sulfide, and then the produced sulfide is re-oxidized. This mechanism has been interpreted in field settings following nitrate injection (Telang et al. 1997).
- Voordouw et al. (2007) suggested that the reason nitrate injection is able to inhibit H_2S production is that NRB generate nitrite in their metabolism and the nitrite is the effective agent that impairs the growth of SRB. (Nitrite is a strong and specific inhibitor of the SRB enzyme responsible for sulfide generation, whereas nitrate does not inhibit SRB.) In this scenario, it would make sense to inject the active ingredient—namely, the nitrite ion. However, nitrite is highly corrosive.

It is possible that some or all of these mechanisms are at play simultaneously and occurring to varying degrees in different parts of the reservoir. What is evident is that the sulfur and nitrogen cycles are closely linked. The balance between these cycles will influence the results and, hence, the effectiveness of nitrate injection in any given system.

However, of these mechanisms, the first two seem to be those most commonly invoked as the mechanisms at play. Dolfing and Hubert (2017) offer some insight into these mechanisms. They suggest that the alternative nitrate-reduction mechanisms can be assessed by means of evaluating the thermodynamics of the difference between the sulfate-driven oxidation of acetate (or other SRB organic electron donors) and the oxidation of sulfide to sulfate. This indicates that, with acetate as the organic electron donor, sulfate reduction to sulfide is always more energetically favorable than the reverse reaction under realistic oilfield conditions. This therefore suggests that acetate would be a more favorable electron donor than sulfide, for example, for NRB and, hence, in a nitrate-based souring-control context. Consequently, nitrate/sulfate competition seems a more likely souring-control mechanism than nitrate-driven sulfide oxidation, with nitrate reduction fueled by acetate and other organic compounds as the pathway of choice.

However, sulfide is not necessarily fully oxidized to sulfate under all conditions. The incomplete oxidation of sulfide to elemental sulfur by means of SONRB can also be the energetically more favorable outcome envisaged, especially under conditions where nitrate volumes might be limited and/or when acetate concentrations are low.

In a simulation exercise, the effects of nitrate injection were modeled on the basis of results of sandpack experiments (Haghshenas et al. 2012), but using the assumption that nitrite is the active SRB inhibitory agent. It found that with nitrate applied at a dosage rate of 300 mg/L, the generated H_2S concentrations were reduced by two orders of magnitude.

A successful nitrate-injection program took place in Denmark's Halfdan Field (Larsen et al. 2004). This case of reservoir souring is rather unusual in that the operator believes that the souring initiates from the production side because small amounts of H_2S are often measured in the produced gas (0.2 to 20 ppm, corresponding to < 2 kg H_2S per day per well) when new wells are brought on stream. Thus, the H_2S is believed to originate from SRB introduced through the use of seawater-based fluids during drilling, completion, and stimulation operations.

Continuous nitrate injection (injection was approximately 9 km away at the Dan Field, where the water-injection facilities are located) was used at a target rate of 100 to 150 mg/L sodium nitrate product (40% w/w), which corresponds to 40 to 60 mg/L of nitrate ion. Measurements at Halfdan suggest some consumption of nitrate in transit from the injection location to the field and also indicate that small amounts of nitrite ion are generated. The backflow of injection wells suggested some depletion of the injected nitrate-ion concentrations and a complete absence of H_2S, suggesting an effective treatment.

Furthermore, molecular bacterial analysis of produced fluids in three wells that suffered rapid seawater breakthrough indicated very low numbers of SRB (maximum of 0.4 per mL) but much higher numbers of NRB (maximum of 15,000 per mL); some depletion of nitrate ion was recorded after the transit through the reservoir, but a positive recording was still taken. Similarly, pigging of the flowline between the injection point and the field indicated high counts of mesophilic NRB (> 140,000 per mL), while SRB were found at very low levels, only 150 per mL.

A comparison between the production performance at seawater breakthrough in Well HDA-07 (**Fig. 10**), where a benefit from the nitrate injection is expected, and that of Well MFA-18 (**Fig. 11**), where no nitrate-injection support occurs, suggests that the nitrate injection had a significant beneficial effect.

These data suggest a successful nitrate application. Nitrate has also been injected in Nigeria's Bonga Field, with a dosage rate of 50 mg/L of active nitrate ion used

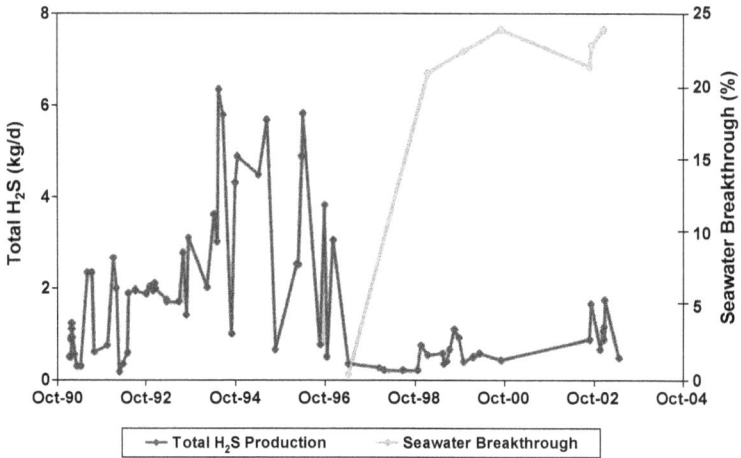

Fig. 10—HDA-07 sulfide production before and after seawater breakthrough (after Larsen et al. 2004).

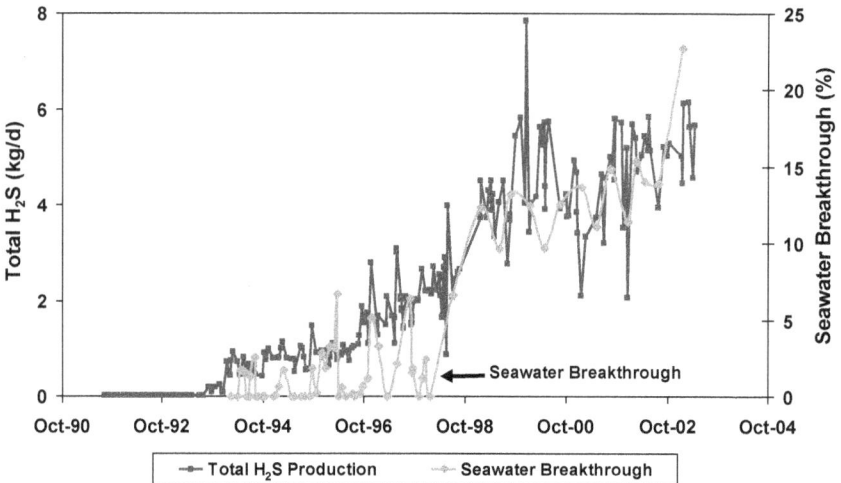

Fig. 11—MFA-18 sulfide production before and after seawater breakthrough (after Larsen et al. 2004).

(Kuijvenhoven et al. 2006), which would normally be considered a reasonable rate. This was expected to reduce the H_2S production potential to a maximum of 50 ppm H_2S, but 5 ppm was expected to be a more likely H_2S level. As far as is known, this field is still producing without H_2S after nearly 15 years of injection.

In the early 1990s, the Gullfaks A platform was suffering from severe reservoir souring. Single wells had several thousand ppm of H_2S in the produced gas, corresponding to approximately 35 mg/L of H_2S in the produced water. The reduction in H_2S production seen after nitrate treatment of the injected seawater corresponded well with the results obtained from laboratory experiments (Sunde et al. 2004). The reduction in H_2S concentrations in produced water at Gullfaks C after nitrate injection began is shown in **Fig. 12,** in addition to the increasing trend that was expected if nitrate injection had not been applied.

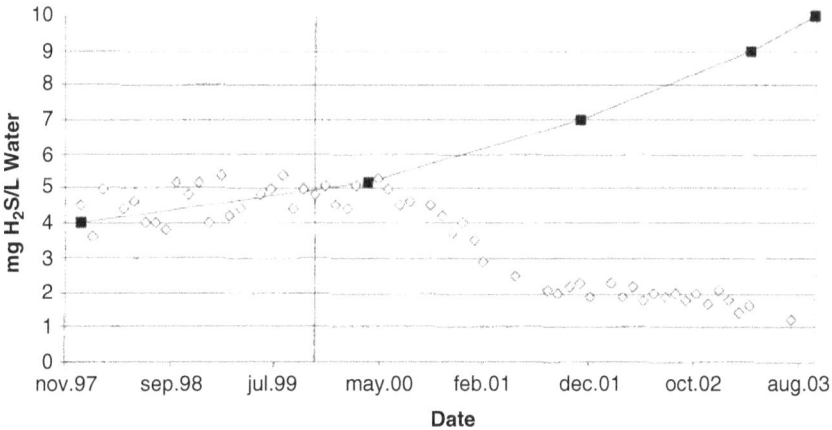

Fig. 12—Measured H_2S concentrations in water at Gullfaks C before and after nitrate injection, compared with theoretical trends if no nitrate had been injected (after Sunde et al. 2004).

Because this was a remedial application, it would not have been expected to completely remove the potential for H_2S production, although it does exhibit an impressive trend compared to the original trend. Furthermore, there was also a reduction in corrosion, which was also observed at a nearby application at Veslefrikk (Thorstenson et al. 2002). Note, however, that although this was claimed as evidence of a successful nitrate application, a similar trend in produced H_2S would have been observed if both biofilm and mixing-zone souring were occurring but the sample timing was such that the spike from biofilm souring had not yet occurred. This explanation might have increased credibility because the positive trend in H_2S trends was not experienced in all wells.

An additional factor that might be expected to critically influence the success of a nitrate-injection scheme is the reservoir-souring mechanism and the question of whether SRB are generating H_2S throughout the reservoir or only in a region near the injector location. NRB are generally expected to grow close to the injection wells because this is the source of the injected nitrate. However, for fields with low temperatures that are able to facilitate bacterial growth at deep reservoir conditions, SRB activity might be driven deeper into the reservoir. For the nitrate injection to be effective, extremely high nitrate-injection rates would be required to ensure that all the food available within the reservoir is consumed by NRB at the expense of all SRB activity. This observation therefore suggests that it would be impractical to inject a sufficient amount of nitrate to successfully suppress H_2S generation for mixing-zone souring (Jack et al. 2009) and that the effectiveness of nitrate injection can be ensured only for higher-temperature reservoirs where biofilm souring prevails.

In the case of the Malaysian MY01 Field, in a reservoir with a temperature of no more than 55°C, reservoir souring of up to 50 to 100 ppm H_2S has been encountered despite a nitrate-injection program (Zhu et al. 2016). Furthermore, a souring modeling exercise in this field has suggested that the observed level of souring is just as high as that expected if nitrate had not been injected.

Although nitrate injection is expected to have some positive effects for most scenarios, because of the uncertainties, there can be no assurances regarding its overall effectiveness unless the reservoir temperature is higher than 70 to 80°C.

One concern regarding nitrate injection is associated with cases where nitrate availability might be limited. Such a scenario might occur when a water-injection plant is closed down for an extended period. In such an event, there is the potential that some of the bacteria species that have been using the nitrate in their metabolic pathways might have the flexibility to start using sulfate when the nitrate is absent. This scenario would reinitiate a souring problem. This possibility suggests that if injection is planned to be closed in for any extended period, it might be prudent to apply a higher dosage just before shut-in and to apply a shock dose to the well just before its closure.

There is likely another important limitation regarding the application of nitrate injection. Many waterfloods inject all the produced-water volumes, using a makeup water source to replace the hydrocarbon off-take. To keep costs down, more often than not those two water sources are mixed. If the makeup water source is seawater, or any water source containing significant amounts of sulfate, there is a reservoir-souring risk to be managed. The problem in this scenario is that produced waters usually contain appreciable amounts of food (e.g., VFAs, BTX) for SRB. Mixing such a water source with a water source rich in sulfate would provide the perfect environment for SRB activity. Consequently, it seems highly unlikely that sufficient nitrate could be effectively, or economically, injected to suppress H_2S production in mixed seawater/produced-water injection systems.

There is an additional issue where nitrate injection is applied in PWRI systems. Although corrosion has not been observed to be an issue with seawater-injection systems, it is sometimes a problem in PWRI systems (Stott 2012). These corrosion effects seem to be system specific; in mild cases, increased corrosion rates appear to be able to be controlled through a corrosion inhibition program, but in extreme cases, corrosion rates of up to 1 mm/yr have been observed. Nitrite ion might be responsible for these problems, although in some cases, additional problems can be attributed to the formation of reduced sulfur species, including corrosive elemental sulfur.

It has been suggested that corrosion effects are experienced in cases where nitrate is only partially reduced (to nitrite) because this tends to result in a partial oxidation of sulfide and could result in the generation of elemental sulfur (Dinning 2011). This indicates SONRB play a role in corrosion effects. Dinning (2011) also suggests that the nitrate/organic-carbon ratio is important and that corrosion rates are lower when this ratio is less than 1. However, that indicates that a choice must be made between achieving souring control and corrosion control.

Corrosion was experienced during the PWRI trial performed at the Draugen Field in the Norwegian Sea (Vik et al. 2007). During this trial, corrosion rates were monitored downstream of the water-injection pumps in the high-pressure system and were found to increase. Although the nitrate controlled the reservoir-souring risks, the increases in corrosion rates were attributed to the nitrate injection and were found to increase logarithmically when nitrate was being injected. At the same time, a few ppm of nitrite was observed in the produced water. Because the addition of biocide reduced the corrosion to background levels and instantaneously removed the observed nitrite, it was concluded that the corrosion was microbially induced.

Overall, these observations suggest that it can be difficult to guarantee success when applying nitrate in PWRI systems; for critical applications, it might be possible to achieve success without experiencing severe corrosion problems if corrosion inhibition is applied in conjunction with the nitrate treatment.

Nevertheless, there have been successful nitrate applications in PWRI systems. One such successful application occurred in the Glauc C Field in Canada (Arensdorf et al. 2009). This field injects a mixture of produced water and treated water from a municipal wastewater-treatment plant. The blended injection water contains 70 ppm of sulfate and no detectable VFAs. Nitrate was injected at a rate of 150 ppm (as nitrate ion), and significant reductions in H_2S were observed. The response was not, however, uniform, and some isolated wells did not exhibit any response. Indeed, nitrate was not detected in producing wells, suggesting that higher injection rates would be needed to generate a more uniform and more efficient response. Nevertheless, over a 16-month period, the H_2S produced per day was reduced from 3.4 to 0.9 kg/d. The results therefore suggest that nitrate injection could still be a viable option for fields subject to mixing-zone souring provided that VFA levels are low enough to be the limiting factor in the level of reservoir souring.

While recognizing the limitations of nitrate injection as a souring prevention philosophy for PWRI applications because it is impractical to inject sufficient amounts of nitrate such that all oxidizable organics are removed by NRB, it has been suggested that because the SONRB-mediated oxidation of sulfide is faster than the NRB-mediated oxidation of organics, nitrate can still be successfully dosed into the production system to reduce H_2S levels (Voordouw et al. 2011). Any such injection should be applied as near to the producing wells as possible, and such a program could potentially be cost-effective in comparison to other sulfide-removal technologies.

The involvement of SONRB has been suggested as one of the reasons that corrosion can be a major issue when nitrate is used in PWRI systems (Dinning 2011). Although the use of nitrate by SONRB can be represented by the following reaction, it is highly unlikely that this occurs in a single process:

$$8NO_3^- + 3H^+ + 5HS^- \rightarrow 4N_2 + 4H_2O + 5SO_4^{2-}. \quad \dots\dots\dots\dots\dots\dots\dots(13)$$

Similarly, the use of nitrate by NRB almost certainly occurs by means of a number of different reactions, with the ultimate product being nitrogen but passing through a number of intermediate species that could include nitrite ion, nitrous oxide, nitric oxide, and nitrogen dioxide. Consequently, the partial reduction of nitrate could feasibly result in the partial oxidation of sulfide and therefore result in the generation of corrosive sulfur species, such as polysulfides or elemental sulfur.

Therefore, nitrate injection can be most effectively employed in seawater-injection systems where reservoir temperatures are high enough to limit H_2S to the near-wellbore region of injectors. Uncertainty remains as to the appropriate injection rate for such applications, and dosage rates to date have largely been developed on an ad hoc basis. Dosages used seem to vary between different companies and have ranged from approximately 25 to 100 mg/L of active nitrate ion (although higher dosage rates have sometimes been used). An appropriate starting point could be 50 mg/L, with the potential to reduce this dosage level in the longer term. Because the dosing strategy should be aimed at ensuring the complete metabolism of all available carbon sources by NRB, there might be an argument for using higher initial injection rates (up to 200 mg/L?) in cases where the initial formation water contains higher VFA concentrations (say, 500 to 1000 mg/L). Because that local food source might be depleted reasonably quickly, the higher dosage rate could then be reduced to the commonly used injection rates within 6 to 12 months.

Concerns have been raised that nitrate injection could result in an increased biomass presence, raising the possibility of an impairment increase in injectors. However, it is likely that any biofilm would be dispersed over a reasonably large area, and this could explain why negative impacts are not typically observed in field applications. However, the potential for significant biofilm development suggests that the later optimization of injection rates could be valuable, aside from any operating expense benefits. An appropriate time to review the dosing strategy appears to be when seawater breakthrough occurs in the producers. Ideally, the dosing levels will deliver a slight excess of nitrate ion all the way to the producing wells, and the presence of nitrate ion in the producers would then allow for successive reductions in treatment levels.

The preferred injection location for nitrate is near the injection well, with surface bacterial populations controlled through biocide. The injection of nitrate farther upstream, into the topside facilities, should be avoided because nitrate injection promotes bacterial activity, and this could increase the risks associated with biofilm formation (and the associated risks of both corrosion and suspended-solids generation) in the topside process. This could induce significant logistical problems in applying nitrate injection to cases where there are satellite facilities linked to a central water-injection location. If it is not possible to locate the nitrate injection at the satellite facilities, a very robust biocide injection should be applied with a regular pigging program. There could also be a need to assess whether extra nitrate injection is needed because of the potential consumption of nitrate between the injection location and the injection wells.

3.6.3 Perchlorate Injection.
Perchlorate has been proposed as an alternative souring-control agent. Its mode of action is analogous to the nitrate mechanism with its reduction to chlorate (ClO_3^-) and chlorite (ClO_2^-). The advantage of this approach is that in addition to promoting thermodynamically preferred processes relative to sulfate reduction, studies have suggested that high concentrations of (per) chlorate could be directly and specifically inhibitory to microbial sulfate reduction.

The industry nevertheless continues to prefer nitrate-injection programs, probably because the microbes that can readily metabolize (per)chlorate are not naturally abundant because of the scarcity of this ion in nature. The effectiveness of perchlorate and nitrate in inhibiting SRB activity in a medium containing heavy oil has been compared in the Medicine Hat Glauconitic C (MHGC) Field in Canada, which has been injected with nitrate to control souring (Okpala and Voordouw 2018). Using acetate, propionate, and butyrate as electron donors, perchlorate-reducing bacteria (PRB) were obtained in enrichment culture and isolated from MHGC produced waters. In batch experiments with MHGC oil as the electron donor, nitrate was reduced to nitrite and inhibited sulfate reduction, but perchlorate was not reduced and did not inhibit sulfate reduction in the incubations. The enriched and isolated PRB were unable to use heavy-oil components, such as alkylbenzenes, which were readily used by NRB.

These results suggest that perchlorate injection might not be a suitable souring-control strategy, at least for the type of system analyzed in Okpala and Voordouw (2018). This conclusion is supported by the fact that a field deployment of perchlorate injection for reservoir-souring control has yet to be made.

3.6.4 Molybdate. Molybdenum is an important trace element used by microorganisms in the synthesis of enzymes responsible for catalyzing reduction reactions, and it plays a key role in nitrogen, carbon, and sulfur cycles. However, high concentrations of this element can inhibit the growth of SRB because the molybdate ion is a structural analog for the sulfate ion and can therefore be transported into the bacteria and result in the deprivation of otherwise available sulfur-reducing compounds.

There is uncertainty regarding the required molybdate concentrations needed to impair growth. A number of studies have been performed in which this was assessed; molybdate concentrations of between 6.4 and 32 mg/L were found to be effective in one study, whereas 320 mg/L was needed in another (de Jesus et al. 2015). To date, no oil field has used this strategy for reservoir-souring control. The cost and the logistics associated with the high required dosage levels are probably the main reasons for this.

3.6.5 Sulfate Removal. Because the presence of sulfate is key to the phenomenon of reservoir souring, it is evident that in the absence of sulfate, reservoir souring will not occur. It might therefore be expected that any significant reduction in the available sulfate concentration will have a positive effect on the amount of H_2S generated. Reducing the sulfate concentration from a given water supply using sulfate-removal unit (SRU) nanofiltration systems might therefore be an appropriate methodology to control the extent of reservoir souring.

SRUs have been widely used in seawater-injection systems since the late 1980s, but the primary reason has invariably been to control serious sulfate scaling problems. However, because so much of the sulfate is removed, they have also had a beneficial impact on reservoir souring.

The first application of SRU technology where reservoir-souring control was the main driver was at the Ursa Field in the Gulf of Mexico, and the waterflood began following an initial period of primary production (Alkindi et al. 2008). The Ursa direct vertical access wells were completed with P110 and CY110 casing and 13Cr tubing. These materials failed qualification testing for 20 ppm H_2S at the bubblepoint pressure, raising a significant integrity concern for the waterflood. To protect the wells from environmental cracking, the engineers defined an acceptable barrier policy to avoid the need for an expensive changeout of well materials. This involved a robust biocide program and the installation of an SRU to deliver injection water with a sulfate specification of 45 mg/L.

The sulfate-removal process is similar to the reverse-osmosis technology used in desalination plants. The water to be treated is pressurized against a membrane that repels sulfate from its surface while allowing sodium and chloride ions through. The flow of the feedwater across the membrane provides a self-cleaning mechanism as the sulfate is removed by the sweeping action of the concentrate or reject stream across the membrane surface. Two streams are discharged from the unit—a clean stream containing reduced concentrations of sulfate (commonly approximately 20 to 40 mg/L) and a reject stream with enhanced levels of sulfate that can simply be discharged overboard (see *Waterflooding: Facilities and Operations*, another book in this series).

Because the sulfate is not completely removed by means of this technology, H_2S generation would not be expected to be completely stopped. However, despite widespread use of this technology in seawater-injection systems over the years, it is thought that H_2S production has only ever been observed in one case, in the South Arne Field in Denmark (Robinson et al. 2010). This is a chalk reservoir with a reservoir temperature of 115°C; H_2S concentrations of up to 35 ppm have been reported for the most severely impacted well. As a chalk reservoir, it would be expected to be devoid of any mineralogy capable of scavenging H_2S, meaning that although sulfate levels are low, all sulfate could potentially be converted into H_2S.

Reservoir souring has also been encountered in a low-sulfate injection scheme in Argentina (Cavallaro et al. 2005). In this case, the project is a produced-water injection scheme with sulfate levels ranging from 5 to 32 mg/L. Isotope analysis has confirmed the produced H_2S to be biogenically generated. This example further demonstrates that although the level of reservoir souring in low-sulfate injection water can be expected to be limited because of the limited sulfate availability, it cannot be eliminated completely.

These observations suggest that sulfate reduction could be a viable reservoir-souring control strategy provided that the amount of H_2S generated is within the natural scavenging capacity of the reservoir. The inference is that sulfate removal is likely to be a viable reservoir-souring control philosophy for sandstone reservoirs. One limitation to using this technology in certain applications could be its high cost. The equipment has a large footprint and, hence, a high cost, especially in deepwater settings. Nevertheless, the difficulties and costs associated with scale squeezes for such applications have led to its wide usage. Therefore, it is evident that, from a souring perspective, souring problems might need to result in a materially significant cost (as is the case at Ursa) before sulfate reduction is used, with souring control as the sole or primary justification.

3.7 Microbial Recovery Impacts. Microbial enhanced oil recovery (MEOR) is a process in which biological activity is promoted with the intention of mobilizing some oil that would normally be considered residual oil in a conventional waterflood.

Suggestions that the use of microbes can increase oil recovery have been around for more than 50 years, but despite many claims regarding the benefits of such treatments, conclusive evidence of the benefits and the nature of the mechanism have remained elusive.

Generally speaking, it is assumed that for a microbial-based process to generate significant incremental volumes of oil, the reservoir conditions are such that the bacteria are able to thrive throughout the reservoir. These considerations often appear to be missing from MEOR screening. Appropriate considerations should include

- Reservoir temperature: In deeper, hotter reservoirs, bacteria are incapable of surviving the reservoir temperature and are expected to thrive only in a relatively narrow, cooled zone around the injection wellbore. Such scenarios are expected to be incompatible with MEOR. The exact temperature constraints are governed by the types of bacteria, but as a general rule, MEOR is expected to be impossible at reservoir temperatures greater than 80°C, and even at somewhat lower temperatures, the presence of thermophilic bacteria strains is required.
- Reservoir pressure: Because most bacteria are expected to equalize internal and external pressures quickly, reservoir pressure might not be a significant parameter likely to constrain MEOR applicability. However, pressure increases the dissolution of gases such as CO_2, thereby affecting pH, which might therefore indirectly affect growth in some cases.
- Formation water salinity: The sensitivity of bacteria to salinity is very much dependent on the species, with each species having a certain salinity window in which it is happy to grow. Most marine bacteria species can happily tolerate NaCl concentrations of up to 6% w/v, but relatively few hardy species can tolerate much higher concentrations.
- Oxygen: Because oxygen is absent from reservoir environments, any bacterial species involved in MEOR processes must be anaerobic.

Although bacteria can have a window of tolerance for the various factors that influence growth, each environment will select for a community of bacteria able to tolerate the prevailing conditions to varying degrees. Although individual species might tolerate a small number of combined factors, the likelihood of survival decreases with each factor that is added, as does the diversity of microbes that can tolerate that environment.

Some MEOR strategies involve stimulating the indigenous bacteria. However, relieving one factor that is constraining growth (such as through the addition of a carbon source such as molasses or through the addition of an element needed to support metabolism, such as phosphate) might not be sufficient if other factors still limit growth. Another strategy that is used in MEOR processes is to add specific organisms to the reservoir. However, for the introduced organisms to thrive, they must be able to not only tolerate the reservoir conditions but also compete against indigenous microbe communities for the available nutrients. Indeed, the practice of dosing a reservoir with a particular species assumes that the species will grow to deliver the desired outcome. This could be difficult to achieve for a number of reasons:

- The indigenous organisms are expected to be better adapted to in-situ conditions and could therefore be more efficient in growth, so there is a significant likelihood that they will outcompete the added bacteria for nutrients.
- The added organisms could fail to penetrate the formation as a result of attachment to surfaces near the injection point.
- Indigenous organisms could consume the added microbes.

Such practices are highly condition dependent and species dependent, which could be one reason for the limited success of MEOR processes. Most literature on MEOR has considered theoretical mechanisms but has not addressed the practicalities and cost-effectiveness of applications. Furthermore, it is not always evident which mechanism is being advocated for any given MEOR application. A significant number of possible mechanisms by which these processes could deliver benefits have been postulated. These include

- A reduction in interfacial tension by means of microbial surfactants: By reducing the interfacial tension between oil and water, surfactants could help to mobilize some of the residual oil. However, it is likely that a very significant reduction (say, two orders of magnitude) of interfacial tension is needed to achieve any material change in residual oil. Unfortunately, most studies have suggested that the changes in interfacial tension achievable by biosurfactants are too small to deliver material changes to residual oil. Furthermore, there could be a tendency for the biosurfactant to adsorb onto the rock surface, meaning that large volumes of surfactant would be needed to deliver appreciable change.
- Wettability change by bacteria: The Low-Salinity Flooding section notes the potentially important role that wettability plays in the displacement efficiency. Unfortunately, it is difficult to quantify the impact that biosurfactants might have on wettability and, hence, on the changes that would be expected in the relative permeability curves. Indeed, the literature is nearly silent on the topic of the alteration of wettability by bacteria. Conceptually, it is expected that the residual oil saturation of sandstones will not change significantly with wettability for permeabilities that are high enough for microbes to penetrate into the reservoir. Also, wettability effects are likely to be smaller than the simultaneous effects of interfacial-tension reduction. In oil-wet carbonates, a change in wettability resulting from biosurfactant production could theoretically increase oil recovery, but the benefits would be expected to be limited by both the amount of surfactant needed to generate incremental oil and the ability of the surfactant to reach low-permeability rock given constraints on microbe penetration.
- Interfacial rheology change: The formation of a biofilm at the oil/water interface will change the rheology of the interface, potentially providing a mechanism to control mobility and areal sweep in reservoirs by reducing the flow of water through high-permeability zones. (Note that rheological properties are not limited to changes in interfacial tension; there could also be viscoelastic changes in the interface.) One commercial MEOR process uses the attachment of bacteria to the oil interface, and bacteria-induced emulsification, to increase oil recovery. In this process, nutrients are added to the reservoir to modify the surface properties of the indigenous microbes with the intention of transforming them into hydrophobic bacteria. The process suggests that in their transformed state, the microbes have the ability to emulsify residual oil and the emulsified droplets can then block thief zones, thereby increasing oil recovery.
- Conformance improvement by the plugging of high-permeability zones: This mechanism—whereby the biological material generated induces the plugging

of high-permeability zones, thereby inducing conformance-control improvement—has received by far the most attention. Bacteria have been shown to be able to travel through core material with a permeability of 170 md, and in this particular case, the addition of nutrients was able to induce a permeability reduction (Jenneman et al. 1984). Effective treatment requires the formation of a long-lasting flow barrier within the formation. This implies that the biomass material that is formed must be resistant to degradation; otherwise, the treatment would need to be repeated at regular intervals. Such a mechanism might be effective if the plugging effect could be limited to the high-permeability areas of the reservoir. This implies that prescreening would be needed to ensure that target reservoirs are identified where flow is dominated by high-permeability layers and where conformance might be improved by such a treatment. Another key factor for the success of this approach could include the identification of indigenous organisms able to produce a biopolymer material that is sufficiently robust to withstand the reservoir conditions.

- Other mechanisms: Additional mechanisms proposed include the increasing of permeability by organic-acid production; microbial gas production in situ that can displace oil and/or reduce oil viscosity, thereby improving the displacement mobility ratio; and microbial solvent production in situ. However, even if such mechanisms were realistic, any mechanism that requires a change to large volumes of reservoir material is unlikely to be viable.

Despite the difficulties associated with MEOR mechanisms, the literature continues to include reports of periodic interest in these processes. Sunde et al. (1992) suggested that the injection of injection water containing oxygen could enable an MEOR process by stimulating the activity of aerobic oil-degrading bacteria. However, they recognize that the availability of oxygen will be a primary limiting factor for such a process. Even in cases where fully oxygenated seawater is injected, it is inconceivable that oxygen could penetrate the entire reservoir, although it might offer some (free) recovery benefits in those parts of the reservoir it does penetrate. It is believed that aerated water injection associated with nutrient injection has been used in the Norne Field in Norway (Rassenfoss 2011), and it is understood that this scheme was originally expected to deliver 6% more oil than conventional flooding. However, it is unknown if this scheme is still active or if the incremental volumes have been realized.

A successful deployment of a nutrient-based MEOR process has been reported in the Trial Field in Canada (Town et al. 2009), where the average producing water cut in the mature waterflood was approximately 95%. Three weeks after the first batch nutrient treatment in an injector, the producing water cut at an offset producer began to reduce, although this seems to be far too quick of a response for this to be an incremental oil bank. The oil production in the most responsive well gradually increased from 1.4 to 8 m^3/d (**Fig. 13**).

Another application of the same technology claimed to experience a successful deployment, with net production rate increases of up to 30% reported (Zahner et al. 2010), although limitations associated with the testing accuracy suggest uncertainty regarding the benefits for this application. Because this treatment is said to have succeeded by using microbes working at the oil/water interface to mobilize additional oil, it is difficult to understand how, or why, the production response to the treatment was so rapid.

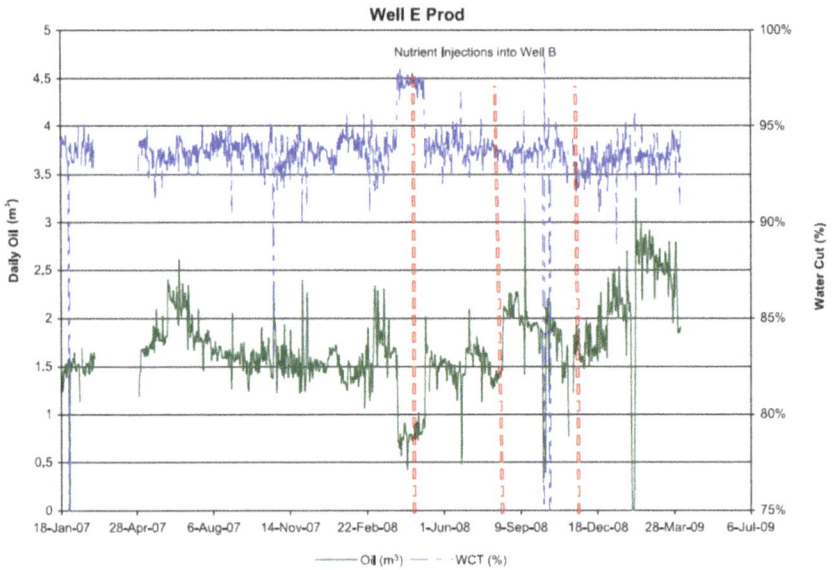

Fig. 13—Oil producer response to a nutrient-based MEOR treatment (Town et al. 2009). WCT = water cut.

Zahner et al. (2011) reviewed more than 100 of these nutrient-based treatments and found the technology is applicable to a wide range of oil densities and is also applicable at high temperatures (up to 93°C) and with high formation-water salinities (up to 140,000 ppm TDS). However, Zahner et al. (2011) noted that the applications have taken place onshore and in very mature waterfloods with very low net oil rates. These are conditions where there could be significant uncertainty associated with the incremental benefits, and the economics of incremental oil in such an environment might be not only challenging but also difficult to verify, especially given uncertainties in oil-cut measurement that are frequently encountered in this sort of environment.

Overall, applications appear to have been sporadic, and even when pilot studies have claimed success, there rarely seems to be more widespread application in the same field. In a review of field-trial data from around the world, Maudgalya et al. (2007) noted that most trials were unable to explain the mechanics of recovery or demonstrate the method by which the benefits had been calculated by means of post-treatment analysis. Furthermore, the results from some trials were found to be contradictory and the claims from others implausible on the basis of the treatment size. Additionally, Maudgalya et al. (2007) often observed that positive results from laboratory studies could not be replicated in the field (which is a major stumbling block for the technology).

Some commonalities were observed in a number of trials, with many featuring additions of a carbon food source (often molasses or other sugar source) and/or nitrogen or phosphorus fertilizers as inorganic nutrients. Furthermore, Maudgalya et al. (2007)

was able to confirm that bacteria are able to travel through the reservoir to producers but that well spacing was critically important to success because microbes consume nutrients as they progress through the reservoir and this then limits the distance from the injector in which they can thrive. Indeed, "microbe retention" near the injector might limit overall treatment effectiveness (Bryant and Lockhart 2000). None of the studies assessed the implications of this, but it might suggest that significant recovery benefits are realizable only for tight well spacings.

Even when positive results were reported, the trials were mostly conducted in very high water-cut, very low net-rate wells, and although some net gains were often claimed, the small volume-gain increases were found to be unconvincing. Indeed, even if it is assumed that MEOR generates incremental oil, this issue probably explains why the technology has not taken off. High-oil-saturation environments should generate the greatest benefits, but the major oil companies that operate such fields are unlikely to be interested unless, and until, the technology is more widely proved. Applications have therefore been confined to the small-scale operators of mature fields with a greater appetite for risk.

4. Scaling

Mineral scale depositional problems are regularly encountered in oilfield systems, and the management of those issues represents a significant operational cost for many fields (Frenier and Ziauddin 2008). Deposition can occur as a result of mixing two incompatible water compositions or because of changes in temperature, pressure, or pH. Consequently, it should come as no surprise that scaling problems are regularly encountered in waterfloods. The location of scaling problems will be influenced by the flood architecture—whether injection is taking place into the water leg (peripheral injection) or into the oil column (pattern injection). Scaling will also be affected by the extent to which the injected water contacts and mixes with the formation water, so it will be influenced by the overall sweep efficiency. If the sweep is good, the majority of the reservoir will be contacted, and mixing will be much better than it would have been if the injection water had short circuited the majority of the reservoir through a high-permeability streak and broken through very rapidly at the producer. It is reasonable to expect that the location of mixing and the amount of injection water in the produced water will change as the flood progresses, so it is therefore also reasonable to expect that the scaling risks will change with time.

The most common scales found in waterfloods are calcium carbonate, calcium sulfate, strontium sulfate, barium sulfate, and iron sulfide.

The mixing of different water compositions will be an issue when a combined injection-water source is used (unless dedicated wells are used for each water source and separate trains are available for each water source), and mixing will also occur when the injection-water source mixes with formation/connate water in the reservoir. Pressure changes have the most effect in the production system, where a pressure reduction is observed all the way from the producing perforations to the production separator. Similarly, a temperature reduction occurs all the way through the production system. In addition, temperature can be an issue in areas where there is a localized temperature change, such as at heat exchangers or near pumps [e.g., injection pumps or electrical submersible pumps (ESPs) in producers].

The solubility of minerals is invariably lower with lower pressure, and as a result, the scaling risks increase as the fluids pass through the production system. For most minerals, solubility is lower at lower temperatures. Calcium carbonate, however, exhibits an increase in solubility at lower temperatures. This occurs because of the impact of CO_2, which has a lower solubility in water at higher temperatures. This, in turn, has an influence on the equilibrium between calcium carbonate and carbonic acid (H_2CO_3). Any process that increases the amount of CO_2 promotes the production of more H_2CO_3, and these increases shift the equilibrium, meaning that calcium carbonate will dissolve. Consequently, calcium carbonate deposition can be experienced at locations where localized temperature increases are observed.

4.1 Prediction of Scaling Problems. The starting point for a robust analysis of scaling risks is to obtain representative water composition data for both the injection water and the formation water. Although that statement might appear to be self-evident, it might not always prove easy to obtain formation-water composition data in new field developments.

Specialized tools are available that enable downhole water samples to be taken, but care is needed in selecting the zone(s) to be sampled because there could be compositional differences between different intervals. The data are likely to be much more accurate when drilling has taken place with a synthetic mud or an oil-based mud system to minimize contamination of the formation water with drilling fluids.

In cases where water-based mud has been used, it is likely that water from the mud has invaded the formation and mixed with the formation water. It should be possible to back out the composition of the formation water from an analysis of the mixture, but this will inevitably carry a reasonably high degree of uncertainty because the composition of the water from the mud that enters the formation will not be known with any certainty and the relative amounts of the formation water and the water from the mud will have to be estimated by using a tracer. In addition, note that it is always possible that there are incompatibilities between the two water sources, which could result in the depletion of one or more scaling species in the analysis.

When the downhole sample is flashed to ambient conditions, the released gas should also be collected and analyzed because its volume and composition will allow for a prediction of the actual downhole formation-water composition.

An alternative to subsurface sampling is to collect surface samples. This can be performed during a drillstem test, or samples can be taken from actual production, if available. It is possible that both of these samples could be somewhat depleted in some ions if scaling has already occurred between the reservoir and the sampling point.

Before a field has been put on production, it is possible that no water samples are available. In such cases, regional data should be analyzed to decide what might represent a reasonable analog (to be determined by a geologist). If core is available, an alternative is to obtain water from the core by means of spinning, but again, this water could have been contaminated by drilling fluids, and it is possible that critical scaling species such as barium (Ba^{2+}) might have been depleted through deposition.

If seawater is the planned injection-water source, it is possible to design the scaling-control requirements based on an average seawater composition, as prescribed in *ASTM D1141-98* (2003). However, it is far more preferable to measure the composition at the planned location, and this should be a relatively easy thing to accomplish because a drilling rig or platform is usually available at the location. Valid reasons real measurements are preferred over generic data include

- There are some regional variations in global seawater compositions— for example, the sulfate concentration could vary in the range of 2400 to 3200 mg/L. Water compositions in confined seas such as the Caspian Sea will be expected to be fundamentally different from those of seawater in the open oceans.
- While compositional variations in the open oceans might be relatively modest, there could be very significant local variations. For example, if the location is close to a major river outlet, the local salinity could be significantly reduced as a result of the seawater mixing with the river water.
- Water samples will, in any case, be needed to quantify an appropriate seawater-intake depth for the seawater treatment facilities.

A number of thermodynamic scaling models are available in the industry. They calculate the thermodynamic driving force for different scales under a specified set of conditions. Calculations can be performed for a specific water composition as well as for various water mixtures having different compositions (for example, various compositions can be expected at different periods in the life of the waterflood based on the changes in the injection-water/formation-water ratio).

The output for each mineral is commonly expressed in terms of two parameters: saturation level and excess solute.

A number of different terms are used for the saturation level parameter, and although they all describe the same process, there is no consistency in the way that saturation is mathematically handled and the values that represent full saturation. Therefore, it is important that when scaling is described in the literature, there is clarity regarding the methodology used. The terms include "scaling tendency," "scaling index," and "saturation ratio"; this book uses the scaling tendency definition. This parameter is a measure of the saturation of a given mineral and is the ratio of the product of the actual ion concentrations to the value of the product at saturation. Scaling tendencies are therefore essentially saturation ratios. A value of exactly 1 indicates the mineral is exactly saturated at those conditions, values less than 1 indicate the mineral is undersaturated, and values greater than 1 indicate supersaturation.

Thus, the scaling tendency is a measure of the thermodynamic driving force for solids to drop out of solution. A critical limitation is that it indicates nothing regarding the reaction kinetics—that is, the speed with which those deposition reactions will be expected to occur. The kinetics of reaction are governed by a number of factors, including the inherent rates of nucleation and growth reactions in addition to the speed at which the mechanisms that transport ions to the nucleation site(s) occur. These will include the rates of diffusion and convection.

Excess solute is a measure of the severity of scaling to be expected. When solids precipitate out of solution, the scaling tendency of the remaining solution starts to decrease. When the scaling tendency has dropped to 1, all the solids that could thermodynamically precipitate have dropped out of solution (although this is a theoretical limit because as the scaling tendency approaches 1, it becomes more difficult for scale to precipitate for kinetic reasons). The excess solute is defined as the volume of scale, commonly expressed in mg/L of the water, that has the potential to thermodynamically precipitate in dropping the scaling tendency to a value of 1.

A high scaling tendency does not automatically imply that a large volume of solids will be formed (i.e., it does not automatically imply a large excess solute). This is because, in cases where the supersaturation is the result of a high concentration of one of the ions of the mineral but the concentration of the other ion is low, the formation of a small amount of precipitate will rapidly deplete the less abundant ion. Consequently, in this case, precipitation will stop after only a small amount of solids deposition even though the scaling tendency is large.

The fact that reaction kinetics do not feature in the scaling models introduces a degree of uncertainty in the interpretation of the output. In other words, the scaling models do not deliver an unequivocal prediction of scaling likelihood, with the exception of cases where they predict that scaling will not occur. There will be many cases where the models predict a problem that will not manifest in reality.

For cases where the driving force for scale precipitation is very large, the occurrence of scaling can be predicted with confidence, but many cases are less clear. Table 2 provides a guideline suggesting conditions in which scaling can be expected, in practice, for different scale types, in terms of saturation levels and scaling severity, based on field experience. Because these are only guidelines, they carry a degree of uncertainty, and different operators could have different views, based on different experiences, about what represents a suitable starting point for the onset of scaling problems.

Table 2—Scaling guidelines.

Scale	Scaling Tendency	Excess Solute (mg/L)
Calcium carbonate	3–4	200–400
Calcium sulfate	1–1.5	50–200
Barium sulfate	5–10	50–400
Strontium sulfate	2–5	75–200

Operators need to carefully consider the specific conditions of each application to determine the appropriate action, especially in borderline cases. These considerations should take into account the scale type that is predicted to deposit. In cases related to calcium carbonate, a more reactive approach might be adequate because this mineral is readily removed by weak acids, whereas a more proactive, preventative approach might be needed for barium sulfate, which produces very hard scales that are difficult to remove. Another key consideration is the question of where the scales are expected to be a problem. The consequences of scale in the topside pipework might be expected to be somewhat less problematic than scale depositing within the perforations at the producing well.

The models enable nodal analysis to be performed throughout the process and take into account the temperature and pressure conditions at each node. After the first set of calculations have been performed, the models offer the option to perform the analyses at each subsequent node without the scale that has deposited, but they also offer the option to carry the scaling ions. Because of the reaction kinetics, less scale will have formed than was predicted by the model, and as a result, the option in which the scaling ions are carried is usually more frequently used because it represents a worst-case scenario.

The deposition of scale deep within the reservoir can be expected to somewhat reduce the amount of scaling encountered at the wells because this removes scaling ions that would otherwise be transported to the well. However, there is a countereffect that, in some cases, could increase the scaling risk at the producing well. This occurs where mineral dissolution processes take place as the injected water passes through the reservoir, thereby increasing the ion loading of the water. Both processes can be present, and it can be a complex task to try to assess the net effect on well scaling. **Fig. 14** shows calcium-ion concentrations in produced water for a field case relative to a linear mixing line between seawater and formation water (in purple) (Houston et al. 2006). It can be seen that the actual data points (in blue) invariably lie above the mixing line, indicating dissolution, or dolomitization, of calcite in the reservoir.

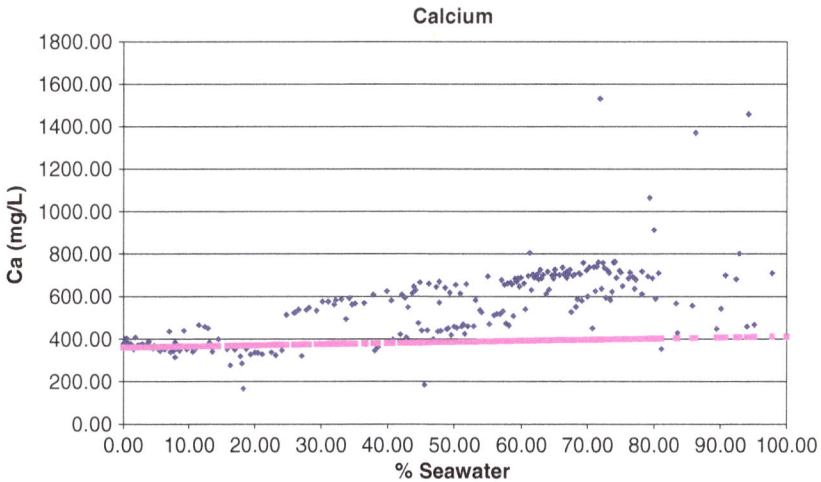

Fig. 14—Calcium (Ca) content in produced water relative to a linear mixing line between chloride-poor seawater and chloride-rich formation water (Houston et al. 2006).

The degree of ion stripping in the reservoir is a function of the degree of mixing between the injection water and the formation water. This means that it can be influenced by factors such as the degree of heterogeneity, reservoir layering, and whether the water has been injected into the water layer (peripheral injection) or into the oil (pattern injection).

Fig. 15 shows an example where reservoir stripping resulting from scale deposition occurs as a result of the mixing of sulfate-rich seawater with barium-rich formation water.

Because the levels of barium-ion concentrations in most formation waters are low and because there are appreciable levels of sulfate in seawater, the deviations of barium concentrations against a straight mixing line are generally very significant, whereas much more modest reductions in sulfate concentrations against the mixing line are observed.

Less commonly, there can also be appreciable levels of sulfate stripping (Mackay et al. 2006). If injection water with a high magnesium/calcium ratio mixes with

Barium and Sulfate Ions vs. Seawater

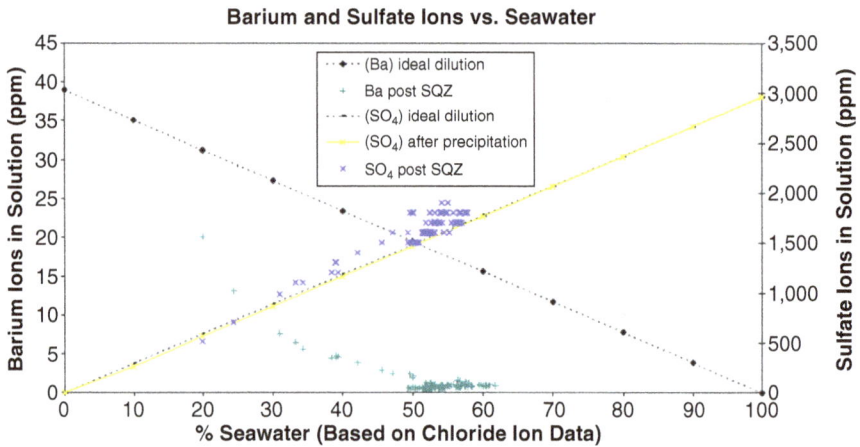

Legend:
- ···◆··· (Ba) ideal dilution
- + Ba post SQZ
- ---■--- (SO₄) ideal dilution
- —— (SO₄) after precipitation
- × SO₄ post SQZ

X-axis: **% Seawater (Based on Chloride Ion Data)**
Left Y-axis: **Barium Ions in Solution (ppm)**
Right Y-axis: **Sulfate Ions in Solution (ppm)**

Fig. 15—Barite stripping (after Jordan and Mackay 2005).

formation water that has a low magnesium/calcium ratio, this can result in magnesium- and calcium-ion exchange with the rock to re-equilibrate. This then results in an increase in the calcium-ion concentration in the water that then induces calcium sulfate precipitation within the reservoir. **Fig. 16** shows an example from the Gyda Field in the Norwegian North Sea. In this case, the barium concentrations lie largely on the mixing line, whereas sulfate breakthrough does not occur until the produced water is composed of approximately 40% seawater (as quantified from chloride-ion concentrations); after that, sulfate levels remain well below the mixing line.

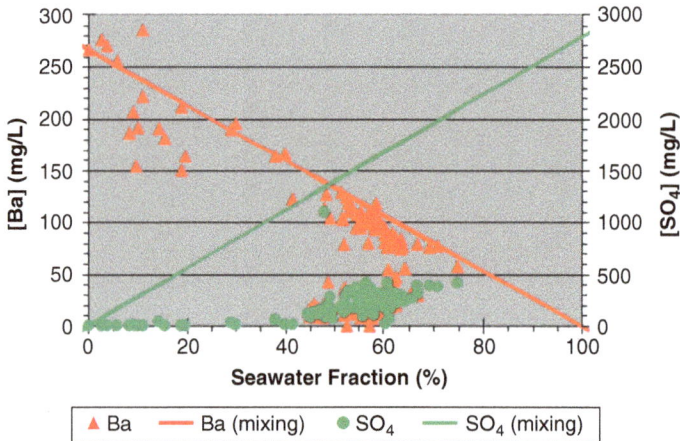

Legend: ▲ Ba   —— Ba (mixing)   ● SO₄   —— SO₄ (mixing)

X-axis: **Seawater Fraction (%)**
Left Y-axis: **[Ba] (mg/L)**
Right Y-axis: **[SO₄] (mg/L)**

Fig. 16—Barium and sulfate concentrations vs. seawater fraction at Gyda (after Mackay et al. 2006).

Fig. 17 shows modeling that takes into account in-situ barium sulfate scale deposition. **Fig. 18** shows the impact when in-situ deposition of calcium sulfate is also

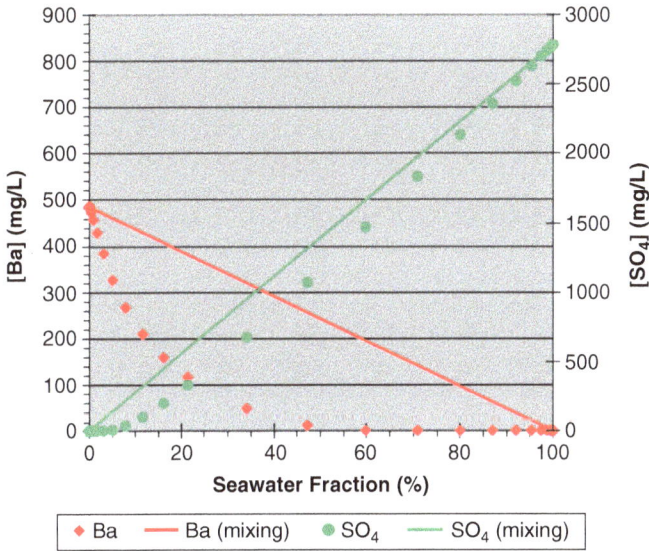

Fig. 17—Modeled barium and sulfate concentrations, taking into account in-situ barium sulfate deposition (after Mackay et al. 2006).

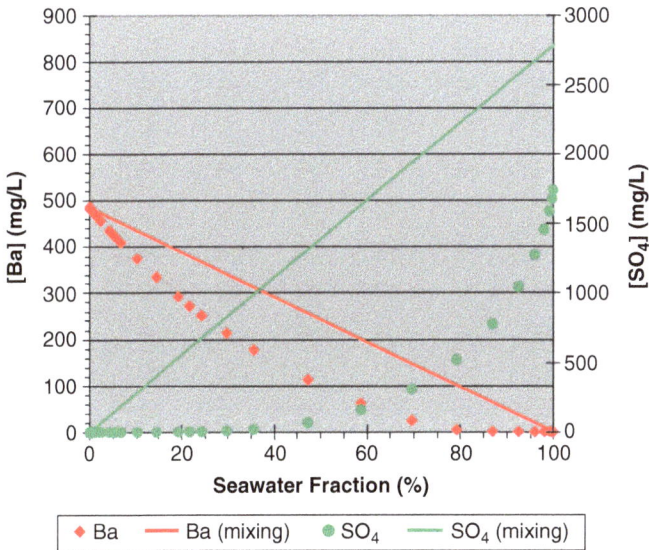

Fig. 18—Modeled barium and sulfate concentrations, taking into account in-situ barium sulfate and calcium sulfate deposition (after Mackay et al. 2006).

considered, and this latter scenario more closely matches the field observed ion concentrations as the seawater fraction develops.

The effects of both the reservoir scaling process and the dissolution process can be captured in more advanced models. These models are more complex to run because they commonly use a reservoir model, but for critical cases, they can be valuable. More discussion of this topic is covered in the next section.

4.1.1 Reservoir Modeling. As previously noted, scaling reactions that occur deep within the reservoir can be expected to deplete scaling ions as fluids are transported toward the producers. It has already been mentioned that in standard scaling assessments, the common approach is to carry the scaling ions so that a worst-case scaling scenario is predicted. For some applications, this could lead to the initiation of a very expensive scale prevention program that is not actually needed. In deepwater settings, for example, such programs can be very expensive. For these critical applications, reservoir modeling exercises can be valuable in providing more accurate predictions that have the potential not only to save in expenditure but also to facilitate production optimization.

Three types of reservoir models might be used to improve scaling prediction: standard finite-difference simulation, streamline simulation, and reactive transport models.

Standard Finite-Difference Simulation. This is the model commonly used for predicting field performance. The reservoir is split into a discrete number of gridblocks. To make feasible predictions within a reasonable time frame, there is usually a limitation on the size of the gridblocks that can be accommodated, and within the gridblocks, the reservoir properties in that region of the reservoir are averaged. The pressure in the cells and the volumetric flow of each phase between cells are calculated, after which the saturation changes are calculated. In principle, this type of model can also be used to track different brines and to assess where they mix. One limitation is that, because there might be a very significant averaging of reservoir properties (it would not be unusual to have a gridblock with dimensions of 100 m), the model tends to underestimate the degree of reservoir heterogeneity, and as a result, it tends to predict that water breakthrough will take longer to occur than it actually does in practice.

A finite-difference simulator will not model scale precipitation reactions and so cannot be used to directly make predictions of brine compositions. However, in cases where field data are available and it is possible to history match the model, more confidence can be attached to predictions of brine composition. There can be little confidence in the prediction of water breakthrough timing, but after it has occurred, more confidence can be attached to predictions of when seawater breakthrough will occur and when scale squeeze treatments should begin.

Streamline Simulation. Streamline-based flow simulation differs from finite-difference calculations in that phase saturations and components are transported along a flow-based grid defined by streamlines rather than moved from cell to cell. This makes streamline simulation more efficient in performing calculations for large models. Higher flow rates result in a greater density of streamlines, and fluid flow can thus be predicted. There is no requirement to update the pressure field with each timestep, reducing the impact of numerical dispersion. This is a major factor when performing brine mixing calculations, affording streamline simulation an advantage in this respect.

Reactive Transport Models. In this modeling, a coupling between the fluid-flow capability of a conventional finite-difference simulation and the chemical simulation capability of an aqueous geochemical model allows a changing formation-water composition and reservoir mineralogy to be studied as functions of time and location in the reservoir. Consequently, the numerical dispersion problems associated with finite-difference simulation are much less of an issue, and this approach is the one most commonly used to study the impacts of reservoir chemical reactions.

These reactive transport models are the ones most commonly used in cases where there is a need to incorporate reservoir dissolution and deposition reactions for a more comprehensive understanding of overall risk.

Although this section focuses primarily on reactive transport modeling in terms of scaling issues, it is important to recognize that such exercises can be valuable in improving the understanding of reservoir-souring processes and low-salinity flooding, as well as a wide range of other processes that are outside the scope of this book.

One reactive transport model, applied to the injection of seawater into a carbonate reservoir, has found that calcite dissolution processes significantly suppress barium sulfate and strontium sulfate deposition processes (Hu and Mackay 2018). However, it has also found that anhydrite (calcium sulfate) deposition is increased because the calcite dissolution provides substantial amounts of calcium such that sulfate ion then becomes the limiting factor. Furthermore, it also results in the deposition of calcium magnesium carbonate because dissolution provides both calcium and carbonate ions and the seawater acts as a source of magnesium ions.

In cases where high-quality produced-water compositional data are available (as they should be), this can be used to history match a reactive transport model. Vazquez et al. (2013) suggested that the results could provide valuable information regarding interwell connectivity, but they would presumably also provide a much improved prediction of the longer-term scaling risks to be expected as the produced seawater fraction increases.

4.2 Scaling Locations and Their Impacts.

4.2.1 Injection System. Scaling problems are less common on the injection side than on the production side. Many produced waters can be saturated with calcium carbonate, and although this will not always induce problems under dynamic injection conditions, it can cause deposition under static, shut-in conditions. The first location where scaling problems are commonly found at the outset of a project is at the injection wellbore where injection water first mixes with the formation water. When these waters are incompatible, scale deposition at the injector perforations can impair the injection well, and the standard practice is to add a scale inhibitor to the injection water when it is first commissioned. After a short period of time, this injection can be halted because all the formation water will have been displaced away from the near-wellbore region and any deposition issues will have been transferred somewhat deeper into the reservoir, where there will be no impact on well injectivity.

Calculating the volume of scale that can be expected in the near-wellbore region confirms the very limited duration of scale problems near injection wells.

A 10,000-B/D injection of seawater into a 100-ft perforated interval of a 20% porosity reservoir would result in a displacement away from the wellbore of 30 ft within a day (Mackay et al. 2003). Assuming that the formation water contains 80 ppm of barium and that all of this was deposited within the 30-ft radius, the scale mass would be less than 5 lbm per 1-ft vertical interval, a very small amount compared to the more than 350,000 lbm per vertical foot of rock mass. Issues such as thermal fracturing can also work to mitigate any near-wellbore scale-deposition problems.

In cases where the injection water is saturated with calcium carbonate, localized heating at the injection pump can also induce deposition that could negatively affect pump performance.

4.2.2 Reservoir. The deposition of scales can occur deep within a reservoir as a result of drops in the overall reservoir pressure if this is not fully maintained during a waterflood. Scales can also be deposited as a result of mixing between the injection water and the formation water as the flood progresses. However, this reservoir scaling is invariably considered nonproblematic because the volume of scale that can deposit at any one location is small relative to the available flow area. This deposition can result in the water eventually produced at the producer being depleted in the ions that are associated with the minerals that precipitate as a result of incompatibility (Li et al. 1996). Indeed, in-situ deposition can be expected to reduce the amount of time that scaling will subsequently be an issue in the production system (Mackay 2002).

Conversely, mineral dissolution is also possible as the injection water transits the reservoir. The ion concentrations in the produced water at Prudhoe Bay in Alaska show trends that can be best understood if the dissolution and precipitation of siderite and calcite in the reservoir are invoked (Li et al. 1996). Because mixing between injection water and formation water might be expected to be greater for waterfloods in which the water is injected below the oil/water contact (peripheral injection), the length of time that any scale squeeze will be required might be more limited for this waterflood architecture (Mackay et al. 2003).

4.2.3 Near-Wellbore Region of Producers. As fluids migrate toward the producing wells, there is the potential for scale deposition in the near-wellbore region. Such scales can deposit as a result of the pressure drawdown that is created in producing the well, but in a waterflood context, the problem is more often associated with the mixing of formation water with injection water in the reservoir close to the well. This deposition has the potential to materially impact well productivity and can even kill the well. Furthermore, such problems can persist for extended periods of time. As such, management of this issue often represents a major challenge (and cost) in waterflooded fields. These issues will be discussed in more detail in the sections that follow.

4.2.4 Producing Wellbore. The mixing of different water sources that can cause scale dropout in the reservoir and near-wellbore region continues in the well, and as a result, scale dropout might continue, especially because the pressure will continue to drop. In the producing well, completion components can be particularly sensitive to scale dropout because only small amounts of scale can have significant consequences.

Components that can be particularly sensitive include inflow control valves, pressure gauges, downhole pumps, and subsurface safety valves. Because reaction kinetics are important, scaling problems tend to manifest in locations where flow is disturbed, and as a result, valves are particularly prone to deposition problems.

In cases where ESPs are used, the motor is cooled by the flow of produced fluids over the housing, which causes a localized heating. Because calcium carbonate solubility is lower at higher temperatures, this can cause scaling problems. Problems can manifest on the housing but might also be located within the body of the pump. ESP run life can be severely compromised by deposition on the pump internals, especially on the rotor tips. In one case, the temperature around the motor was recorded as being 10 to 24°C higher than that of the remaining bulk fluid (Wylde and Fell 2008). This often results in the incorporation of a shroud to increase the fluid flow alongside the motor, thereby improving the cooling and leading to an improvement in ESP run life.

Failure of the safety valve commonly requires a well intervention, and because of the significant consequences in subsea developments, where such interventions are very costly, a particularly proactive scale-control strategy is commonly required. Carbonate scaling increases in severity as the pressure drops, and when carbonate scaling of the safety valve is a potential issue, the risk can be reduced by deeper placement of the safety valve.

Deposition can also occur on the walls of the tubing. This results in a restriction of the flow area that can further increase the pressure drop in the system and will reduce well productivity.

Because calcium carbonate deposition is governed by the equilibrium between CO_2, bicarbonate (HCO_3^-), and carbonate (CO_3^{2-}), calcium carbonate problems first become possible downstream of the bubblepoint. This is because as the pressure drops and the bubblepoint is reached, flashing of CO_2 will be induced. That has the effect of shifting the equilibrium, so this signals the point at which such deposition can begin to occur. In many cases, the bubblepoint pressure will be reached somewhere within the wellbore. Neither the thermodynamic driving force for scaling nor the scaling mass potential increases dramatically at the bubblepoint, so it is likely that the turbulence associated by the release of gas bubbles helps to precipitate the problem.

4.2.5 Wellhead and Gathering System. The next constraints in the process are the wellhead valves, where deposition can again be induced by pressure drops and turbulence. As is the case with deposition in the well tubulars, solid deposition in the gathering system and pipelines will reduce the area available to flow so that it has either a negative impact on production or requires that additional pressure be applied to maintain the flow rate.

Fortunately, scale buildup in this part of the process is normally quite slow because the volumes of scale that can deposit at this stage of the process are usually more modest. Additionally, there is often a large volume of pipework and thus a large available surface area for deposition. On the other hand, the removal of deposits from this part of the process is more difficult. Large amounts of chemical dissolvers could be required, and the temperatures could be too low to facilitate the chemical removal of sulfate scales.

4.2.6 Processing Facilities. As is the case with the induction of scale at the bubblepoint in the wells, CO_2 will also flash off in separator vessels, and this could induce carbonate scaling or other scales that are pH dependent. There could be negative impacts associated with separation efficiency when scales are manifest at this location because the solids might promote emulsion stability; it is also possible that level control valves could suffer failures.

Many topside processes use heat exchangers and, as is the case where temperature changes occur at pumps, calcium carbonate deposition problems can be induced. Scale deposition reduces the heat-transfer efficiency of the exchanger, which then demands a higher temperature of the heater medium, accelerating the problem.

Equipment using very fine filtration, such as SRUs, can be particularly prone to problems when scaling risks are present because fouling of the membranes will be expected. Because divalent ions are impermeable, their concentration will increase close to the membrane surface, and any deposition occurring in that area will also cause problems.

When H_2S is being produced, iron sulfide could potentially form. Because oil is also present, schmoo formation (see Section 4.5.6) is a problem. In some cases, schmoo could remain dispersed as small particles that pass through the subsequent water-treatment part of the process before subsequently agglomerating. It could induce pipeline deposits, and in PWRI systems, this can induce injectivity problems.

4.3 Scale Prevention. A scale management strategy needs to be built that recognizes the inherent uncertainty associated with scale prediction capabilities. Thus, the scale prediction will develop an understanding of only the likelihood that scale precipitation will occur. This likelihood then needs to be balanced against the impacts associated with scale deposition at the expected location. Those impacts can be different for different scale types and will certainly be different for different locations in the process. Thus, a preventative approach is likely to be used in the event that barium sulfate scaling is predicted at the perforations of a subsea well in a deepwater environment. By comparison, calcium carbonate scaling in a more accessible part of the process is more likely to result in the adoption of a wait-and-see approach.

The volume of scale can also play an important role in determining the appropriate strategy. In some specific locations, such as the subsurface safety valve or an inflow control valve, even very small volumes of scale could induce damage, resulting in very expensive consequences.

The difficulty associated with scale control in different locations is based on a combination of accessibility and completion type (Mackay et al. 2002; Mackay et al. 2005a).

Methods that have been applied to prevent, or at least minimize, scaling include

- Minimizing pressure and temperature drops in the system
- Avoiding the mixing of incompatible brine streams
- The removal of at least one scaling component from the system, either chemically or physically
- Changing the pH
- The application of magnetic fields
- The use of scale inhibitors

Magnetic fields (Donaldson and Grimes 1987) have been used periodically but have never become established as a mainstream scaling-control philosophy. This is probably because a cohesive mechanism whereby such a philosophy would work remains elusive, and moreover, there seems to be little understanding of the field strength or the duration of treatment needed to control scaling for any given setting. From the available information, it seems that single-pass systems, which would be required for effectiveness in most oilfield systems, are not often effective and that if such systems are found to be effective, it is usually in the treatment of calcium carbonate scales—it has been claimed that magnetic treatment causes water containing minerals to favor the formation of the more soluble aragonite form of calcium carbonate rather than calcite.

A series of controlled laboratory tests were performed to assess the impact on calcium carbonate and calcium sulfate (gypsum) scaling as supersaturated brines were passed through a 13,500-gauss field created by an electromagnet and through a pair of dummy and magnetic tools for downhole application (Pritchard et al. 2000). In these tests, neither treatment had a significant effect on the amount of scale deposited, the morphology of the scale, or the turbidity of the mixed brines over a wide range of supersaturations.

Overall, it appears that, because of the lack of a comprehensive understanding of how magnetic scale prevention might work and the critical success factors in such a preventative program, the use of magnets presents a risky scale-control option. This is almost certainly why nearly all mainstream scale-control programs use "threshold" scale inhibitors.

Classes of chemicals that fall in this category are able to prevent the formation of solids in circumstances where scaling would be expected on the basis of reaction thermodynamics. They work by preventing nucleation, whereby the proto crystals form but are then either dissolved or disrupted by the action of the inhibitor molecule and/or by adsorption or interaction with the crystal growth site that retards or blocks ongoing growth at the crystal site (Sorbie and Laing 2004). Both mechanisms probably contribute to the action of the inhibitors, but it is thought that smaller phosphonate inhibitors primarily act as growth inhibitors, whereas polymeric inhibitors work primarily as nucleation inhibitors. Because of their mode of action, they are able to be effective at relatively modest injection concentrations—dosage rates of less than 20 mg/L are usually sufficient to control carbonate or sulfate scaling. (Note that there is another class of scale inhibitors, chelating agents, that has a different mode of action such that they need to be applied in stoichiometric ratios and therefore generally entail a much greater cost.)

Seeded growth studies have suggested that it is probably necessary to cover 5 to 25% of the growth sites to prevent seed crystals from growing (Tomson et al. 2003). The surface area covered by a single inhibitor molecule is unknown, but to cover 5 to 25% of the surface area, Tomson et al. (2003) estimate that the inhibitor concentration needs to be between 0.05 and 0.25 mg/m^2. Nucleation studies were conducted with various concentrations of an inhibitor (**Fig. 19**) and showed that no inhibitory effect was observed at concentrations less than 0.14 mg/L, corresponding to approximately 2.7% surface coverage. At concentrations corresponding to a surface coverage between 2.7 and 16%, the logarithm of the nucleation time was linearly related to the inhibitor concentration, and at coverages greater than 16%, the solution was fully inhibited for more than 30 days. The results, which were found

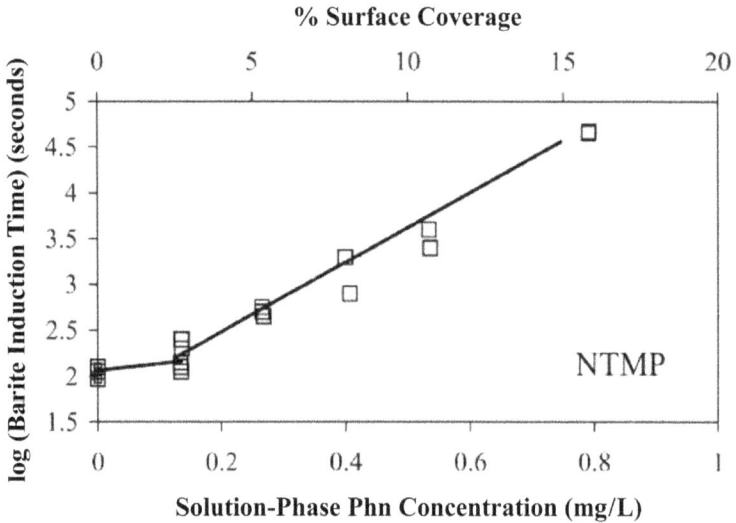

Fig. 19—Induction time vs. inhibitor concentration (Tomson et al. 2003). Phn = phosphonate, NTMP = nitrilotri(methylenephosphonic) acid.

to be surprisingly independent of the mineral type, suggest the possibility that this might provide a means to identify the maximum inhibitor concentration required for this inhibitory mechanism.

There is a wide range of different chemistries that are effective as threshold inhibitors (Kelland 2009), including phosphonates, phosphate esters, polyacid polymers (e.g., polyacrylic acid, polymaleic acid, polyvinylsulfonic acid), and polysulfonates. The appropriate chemistry for any given application needs to take into account the type of scale to be controlled, the water chemistry (many inhibitors do not work well in the presence of iron), the temperature regime, and if there are any compatibility issues with other chemicals used in the process.

The mode of action of threshold inhibitors requires that the chemical be applied upstream of the onset of deposition, and they require continuous injection. For most topside applications, this can be achieved through an injection quill and a small injection pump placed directly into the stream. Where problems are anticipated within the producing well, delivery of the chemical becomes more problematic and costly. One option is to drop sticks containing the inhibitor into the well, which then release the chemical, but there is little control over the applied dosage with this option. Alternatively, the chemical can be dosed into the gas lift system, assuming that this is the lift philosophy used in the field.

If the chemical is dosed into a system serving multiple wells, there is no guarantee that the chemical will be delivered at the same rate as the lift gas to each well, so it is preferable to dose into the annulus at the wellhead. However, that option also has a number of important drawbacks because there is a tendency for the chemical to gradually fill the annulus and slug into the well when the level reaches the gas lift valve, making the dosing somewhat inconsistent. Furthermore, there can be some flashing of the inhibitor solvent, which could result in a gunking of the chemical that can cause plugging of the gas lift valve.

Therefore, the best—but not the least costly—option is often to dose the chemical by means of a capillary injection line, which is usually strapped to the outside of the tubing.

4.3.1 Scale Squeeze. As the location of the scaling onset moves farther upstream, the problems associated with dosing of the inhibitor become yet more complex. Scaling in the perforations and the near-wellbore region of the producer can be suffered for a reasonably long period of time when injection water breaks through and mixes in this region with the formation water. This problem requires the inhibitor to be dosed into the reservoir. Unfortunately, it is not really feasible to dose the chemical into the injection system to protect against scaling problems at this location. One reason for this is that there is no control over the route taken by the water through the reservoir, but far more importantly, adsorption of the chemical on the formation would prevent it from reaching all the way to the producing well.

It is now possible to combine sand control and scale inhibition into a single program by impregnating a ceramic proppant with a chemical scale inhibitor and then pumping this proppant as part of the gravel-packing operation (Selle et al. 2010). This option is sometimes useful, but it does suffer from two key limitations. First, if there is a significant formation water bank produced ahead of the injection-water breakthrough, the scale inhibitor could be washed out of the completion before it is actually needed. Second, the mixing of injected water with formation water could be taking place in the near-wellbore region, slightly upstream of the treatment location, and thus, this option might not be effective; this was a problem experienced in Selle et al. (2010). As such, these factors suggest that such applications might be limited to the treatment of in-well scaling problems.

As a result, scale-inhibitor control for problems in the near-wellbore region requires the inhibitor to be placed into the formation by means of a scale squeeze treatment. Because chemical inhibitors usually have good reach, it can be expected that this dosing will also protect against scaling problems within the well, provided that the dosage coming back into the well is adequate.

In a squeeze treatment, the producer is closed in and the inhibitor is squeezed into the reservoir. Much of the inhibitor is absorbed by the reservoir rock, and after a short shut-in period, when the well is brought back on stream, some of the inhibitor tends to be immediately back produced. However, as production continues, so does the desorption of the inhibitor, and the concentration of the inhibitor being back produced then protects the well. The concentrations of inhibitor being back produced need to be monitored because when the levels drop below the concentration needed to maintain control [the minimum inhibitor concentration (MIC)], the well will need to be retreated.

Despite the costs associated with scale squeeze treatments in producing wells, the negative consequences of failing to control significant barium sulfate and/or strontium sulfate scaling problems are often more significant. Tjomsland et al. (2001) calculated that the cost associated with scale squeeze treatments and other aspects of scale prevention at the Veslefrikk Field, offshore Norway, from the onset of water breakthrough in 1992 until 1999 was USD 6.3 million. Their analysis showed that without the scale-control program, the field would have produced 9×10^6 m^3 less oil, indicating that the treatments carried a value of USD 1.1 billion. Assuming constant reserves, the accelerated production represents an incremental USD 320 million in the lifetime net present value of the field.

Fig. 20 illustrates an example of barium sulfate scaling prediction at the perforations for various ratios of injected seawater to the coproduced formation water. The shape of this plot is typically observed in seawater-injection schemes. It shows that the maximum mass of scale is encountered at low seawater fractions (i.e., early after water breakthrough). However, the maximum scaling tendency invariably occurs at higher seawater fractions, usually in the range of 50 to 60%. This is because the degree of supersaturation is higher at these seawater fractions, even though the mass of scale precipitating from a given volume of water is lower. These trends mean that squeeze treatments might need to be repeated a number of times before the flood is mature enough that the fraction of seawater breakthrough is high enough that the scaling risks are sufficiently reduced. The timing of this occurrence will be a function of the scaling severity for each application. The monitoring of scaling ions and diminishing levels of returning scale inhibitor are needed to quantify the optimal timing of retreatment. For cases where serious scaling problems associated with sulfate scaling are predicted, it might be necessary to consider a pre-emptive scale squeeze treatment—that is, perform a squeeze before any scaling is expected, which usually means before water breakthrough has occurred.

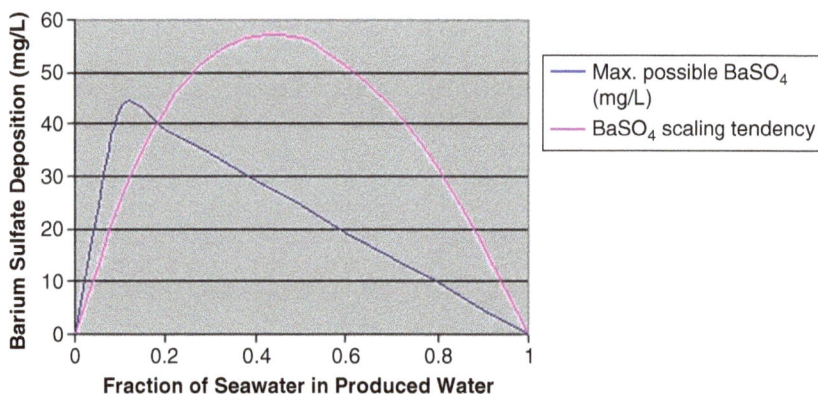

Fig. 20—Scaling risk at producers for various fractions of seawater. BaSO$_4$ = barium sulfate.

In a scale squeeze, the following stages of treatment are normally applied:

- A preflush (sometimes called the spearhead) aims to clean up the formation and enhances water-wetting, therefore promoting the inhibitor adsorption. It will also control any incompatibility with the formation water, if there is any. The preflush incorporates a low concentration of surfactant and might include a mutual solvent to help clean up the formation. Vazquez et al. (2009) suggest that a mutual-solvent preflush could have the potential to extend squeeze life, which could be particularly attractive in subsea settings, although a sizable preflush volume might be needed to achieve such an effect. The preflush is often primarily based on a potassium chloride solution or seawater.
- The main flush incorporates the bulk of the inhibitor (although a small amount might also be included in the preflush). The inhibitor is normally pumped in a concentration ranging from 5 to 20%.

- An overflush is designed to displace the inhibitor away from the wellbore to maximize the subsequent squeeze lifetime (i.e., to protect the well against scaling for the maximum amount of time for a given volume of produced water). Ideally, a squeeze should be designed to last for at least a year. The addition of a squeeze-life-enhancing product to the overflush could also be useful. Sutherland and Jordan (2016) suggest that these enhancers work by means of slowing the initial desorption of the inhibitor from the formation mineral surfaces rather than by aiding the retention of additional chemical.

Two principal, different mechanisms are used to achieve initial retention of the inhibitor in the reservoir. The adsorption process is a physical adsorption/desorption interaction of the inhibitor with the surfaces of the reservoir minerals (Jordan et al. 1994; Jordan et al. 1995). The adsorption is thought to occur by means of electrostatic and van der Waals interactions between the inhibitor and the mineral surface. Not surprisingly, the reservoir mineralogy—specifically the clay types and their levels of abundance—plays a key role in this process. The second retention mechanism is based on precipitation (Rabaioli and Lockhart 1995; Jarrahian et al. 2019). In this process, the divalent-ion concentration plays an important role in controlling the precipitation onto the rock surface after adsorption has taken place, and it is thought that precipitation occurs as a result of the formation of a sparingly soluble calcium-inhibitor complex. The precipitation option might offer advantages in that squeeze lifetimes could be extended.

Additional retention mechanisms are available using droplet entrapment by means of a macroemulsion (Collins et al. 2001) as well as a microemulsion formation (Wat et al. 1999).

After pumping the treatment, the producing well is typically shut in for 12 to 24 hours to allow adequate time for the inhibitor retention mechanisms to be completed.

The selection of an appropriate inhibitor is an important consideration for a scale squeeze. The performance of inhibitors is commonly evaluated by static and dynamic tests, and these tests are also used to assess the required MIC to inhibit the scaling tendency. A coreflood test is usually a key element in the chemical evaluation program before a first squeeze in a reservoir is conducted. It can be used to evaluate different inhibitors and to assess the expected squeeze lifetime. Formation-damage and chemical-return data are needed to enable treatment volumes to be calculated by deriving an isotherm from the coreflood data.

When a chemical has been selected, a model can then be used to optimize the treatment design, including treatment volumes and placement considerations (Mackay and Jordan 2003).

Compatibility and stability testing of the scale inhibitor is also required because the scale inhibitor must be compatible with all materials and fluids for much longer periods, and at higher temperatures, than is the case for conventional scale-inhibitor deployments. Therefore, a qualification program for the scale inhibitor should be considered a requirement to ensure an acceptable subsequent performance (Halvorsen et al. 2009).

There also needs to be consideration of how long the scaling problems will persist in the near-wellbore region to understand the extent of the period that squeeze treatments will be required. In a waterflood context, the problems will only begin when the injection water breaks through in the producing well. If the consequences of the scale are severe enough, there could be a need for pre-emptive treatments so that protection is already in place when injection-water breakthrough occurs.

This approach is difficult to get right because any formation-water production before injection-water breakthrough will have already started to remove the scale inhibitor from the squeeze. In extreme cases, it is possible that the squeeze inhibitor could be washed out completely before injection-water breakthrough occurs. This problem is magnified by the uncertainty associated with injection-water breakthrough timing because geological complexity is nearly always underestimated, which tends to mean that water breakthrough occurs before the predicted timing.

Diversion. In its simplest form, a scale squeeze is bullheaded into the formation. However, the aim is to prevent scale deposition in the near-wellbore region and to maintain maximum productivity of that well. This therefore requires that the inhibitor contact all parts of the near-wellbore region where injection water can mix with formation water and deposit scale. Nearly all waterfloods suffer from suboptimal sweep because heterogeneity results in the injected water flowing through preferential flow pathways. A similar phenomenon often occurs in the squeeze operation unless attempts are made to divert the injected fluids so that the maximum area of the near-wellbore region is contacted by the inhibitor. It is for this reason that many squeeze programs must use some form of diversion.

The preferential flow pathways that tend to develop in waterfloods can cause an additional complication for scale squeeze treatments. The nonuniformity of flow tends to result in the development of pressure differentials, and wells in which this occurs can suffer from crossflow problems when the soak period begins after placement of the treatment. This is likely to result in a shorter squeeze lifetime, and there could be suboptimal scale control even though inhibitor is detected in the produced water. This is not an easy problem to solve; viscosifying the treatment is probably the most effective remedy.

An additional, important reason that diversion is required is that, as time goes by, the amount of inhibitor backflowing from the squeeze treatment gradually reduces until an MIC is reached, at which point the inhibition program is no longer effective. At that point, an additional squeeze treatment might be needed. However, if diversion has not been used, it is possible that significant portions of the well will be unprotected for an appreciable amount of time before the MIC is reached.

The problem is essentially similar to the one that faces many stimulation programs. Diversion is becoming increasingly challenging because of the more complex completions being used today; the remote location of many wells in subsea settings adds a severe complication to the design. Factors that influence the adequate placement of the inhibitor across the perforation interval by means of diversion and that tend to make this difficult include

- Production interval length: Many waterfloods now use horizontal wells, which tends to make placement more difficult. In the simplest cases, it might be possible to achieve placement simply by pumping at high rates.
- Completion: Sand-control philosophies and the options to improve flood conformance within these completions can make placement particularly challenging.
- Reservoir properties: In heterogeneous reservoirs, differences in permeability clearly influence where the injected chemical will preferentially flow. However,

this will be further complicated by the fact that heterogeneity results in the development of pressure differentials; furthermore, some zones will have been preferentially swept, so there will be no uniformity in the saturations along the wellbore. This will further complicate placement.

- Fluid density: In horizontal wells, the density difference between the injected fluids and the in-situ fluids can cause some segregation. During the shut-in period, there could then be some slumping of the water-based inhibitor fluids under the forces of gravity.

Both mechanical- and chemical-diversion (Jordan and Mackay 2009) options can be used. Because well access is not always readily available when a scale squeeze is needed, there might sometimes be a tendency to avoid the mechanical-diversion options. If mechanical diversion is required, one obvious option is to use coiled tubing to place the treatment.

One chemical option that has been widely used is wax as a diverting agent (Ravenscroft et al. 1996) because the wax loses its mechanical strength as fluids are back flowed, and as it melts, it becomes completely miscible with the crude oil. Foams are rarely a viable option because the high reservoir temperatures tend to render them ineffective, although they might be an option in lower-temperature reservoirs. Another chemical-diversion option is to lightly viscosify the entire treatment using gel pills or particulates. For a squeeze treatment in the Nelson Field, UK North Sea, each stage of the treatment was viscosified using xanthan-based biopolymers (James et al. 2005). An important learning associated with this treatment was that some process problems were encountered when the well was brought back on stream because an insufficient shut-in time was allowed to ensure the complete breaking of the gel. Emulsification of the inhibitor package is another option that is sometimes used to achieve the required diversion (Jordan et al. 2003; Feasey et al. 2004).

In highlighting the importance of placement in relation to both squeeze lifetime and inhibition effectiveness, it is evident that the squeeze will need to be pumped under matrix injection conditions because any fracturing will have a negative impact on placement. Therefore, rigorous definition and control of injection pressure becomes an important factor for such treatments.

Squeeze in Subsea Systems. Scale squeezes in wells where direct vertical access (DVA) is available are much simpler to design, and much less expensive to pump, than subsea wells. When pumping squeezes in subsea wells, the deployment costs are likely to far exceed the chemical costs, which differentiates these treatments from applications in DVA wells.

It has previously been noted that the scale management philosophy should be based on risks and consequences. In deepwater subsea systems, for example, a much more conservative approach to scale control is likely to be taken because the consequences of problems are much more serious. This implies that scale control should be more rigorously applied in such systems or that an alternative should be selected.

For a case in Angola, the injection was successfully achieved by means of a gas lift system (Poggesi et al. 2001), but this is a difficult option and one that is not easy to get right. One advantage of this option is that the gas lift line is usually quite clean, but care needs to be taken to manage hydrate risks. For the Kestrel Field in the

North Sea (Kelly et al. 2005), both a scale-dissolver treatment and a scale squeeze were pumped where two producers were connected by means of a daisy-chain arrangement. After placement of the treatment, the risk of hydrate formation in the second well when gas lift was restored was managed by injecting methanol into the tree of that second well by means of a configuration that forced it through the cross-over valve and into the gas lift line between the wells. The volume pumped was sufficient to displace the line volume, and at the receiving well, the methanol flowed into the tree through the crossover valve and then, with the choke cracked open, into the production line. Thus, with both trees and the gas lift line filled with methanol, the hydrate risks at startup were minimized. Then, to further limit any risks associated with liquid accumulations remaining in low points of the line, a pill comprising 70% ethylene glycol and 30% water was injected into the line as a hydrate-inhibiting spacer.

The challenges in achieving adequate squeeze placement have become increasingly difficult in recent years as field developments have expanded into deepwater settings that are most commonly developed using subsea systems. The challenges include very high costs associated with well access, difficulties in getting a squeeze treatment down to the reservoir and then placing it where it is needed, high levels of deferment, and difficulties in monitoring water chemistries to optimize squeeze timing. High frictional pressure losses are often a problem when squeeze treatments are pumped from the host in a subsea system.

The default option for squeeze treatments in the North Sea has been to use the production flowline. Although this can be a low-operating-expense option, it will typically entail very significant associated production deferment. Furthermore, as the squeeze is bullheaded back into the formation, this option has a very high risk of inducing formation damage because material accumulating in the flowline could be removed by the squeeze; thus, the cleanliness of the flowline is critically important to avoid scales, waxes, corrosion products, and others from being swept back into the well.

For one successful squeeze campaign in a deepwater setting, in Brazil, squeezes had to be bullheaded from the host through the subsea system as a result of rig unavailability (Bogaert et al. 2006). A high barium sulfate scaling tendency demanded that squeezes be used even at low water cuts after seawater breakthrough had occurred. Initially, hybrid squeezes were applied to limit the volume of water pumped. A hybrid squeeze uses an aqueous main flush but an organic flush for either the preflush or the post-flush. The treatments used a mutual solvent, an aqueous main flush, and finally a diesel overflush. In a later campaign, the diesel overflush was replaced with seawater because of significant issues with diesel injectivity. The flowlines were cleaned by means of pigging, and hot oil was circulated until returns were "clean." The oil rates before and after the treatments from both campaigns are shown in **Fig. 21,** demonstrating that the treatments were achieved without damaging the wells.

Another (expensive) option is to deploy from a rig using coiled tubing connected to the tree by a remotely operated vehicle. An oversized methanol dosage line was used to place treatments at one location in Norway (Klepaker et al. 2002).

An additional complication in the management of scaling risks for subsea developments is that surveillance is usually a fundamental requirement to monitor returning inhibitor levels so that it can be determined when the well needs to be retreated.

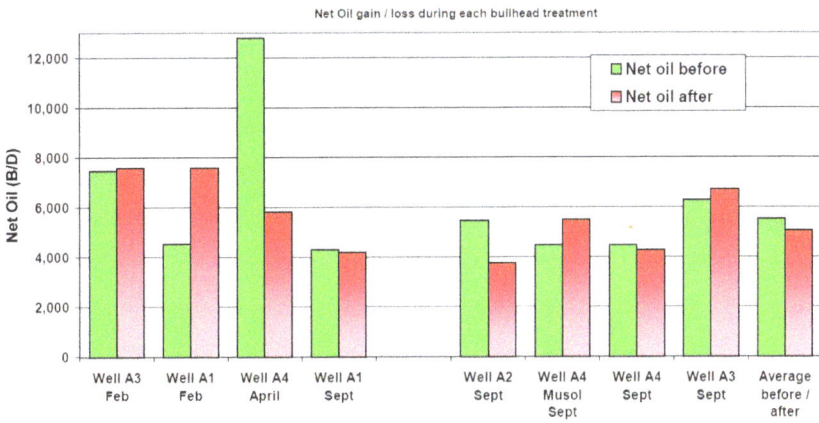

Fig. 21—Oil rates before and after squeeze treatment (Bogaert et al. 2006).

Because well fluids are normally commingled in the subsea line, the sampling and well-testing capabilities in subsea systems are usually inadequate. These problems often result in scale problems being managed by means of sulfate removal (see Section 4.3.2) rather than scale squeeze. However, one novel solution entailed the use of different inhibitors for different wells (Heath et al. 2014). This solution facilitated the pumping of squeeze treatments in subsea wells from the host, and the inhibitor returns from each of the three wells treated were then quantified from the combined produced fluids by means of inductively coupled plasma mass spectroscopy analysis, which offers high precision, accuracy, and sensitivity.

Resqueeze Timing and Optimization. On the basis of the scaling tendencies in Fig. 20, for fields with significant sulfate-scaling tendencies, it can be observed that the scaling risks can persist for a very long time. As a result, well retreatments are commonly required. In one case, the use of a tracer injection (potassium chloride was chosen because of a large contrast between the potassium concentration in the injection and formation brines) was able to provide valuable information regarding the extent of crossflow in the reservoir (Vazquez et al. 2014). The potassium chloride brine was injected into the producing well before shut-in to allow crossflow to initiate. When the well was brought back on stream, the returning potassium concentrations were recorded and compared with those expected based on a history-matched near-wellbore model. The updated model was then used to improve the subsequent treatment design. Production logging tools are an alternative that might provide a similar type of data that can be used for future squeeze design (Jordan 2019).

There is then the question of the time at which a resqueeze treatment should be performed. A number of different, complementary surveillance data are used to quantify this (Ramstad et al. 2009). There are limitations with each of the individual surveillance elements, so they should be assessed in combination to determine the appropriate time for retreatment.

First, the relevant scaling ions are monitored to detect any changes to their concentrations that might suggest an increase in deposition. If the analysis is not immediately performed on-site, it is important that samples be properly preserved (e.g., acidization, addition of a chelant) such that changes to the scaling ions do not occur over extended periods of time as a result of deposition before the analysis takes place. The water-cut development is also relevant because any unexpected increases in the producing water cut will be expected to decrease the anticipated squeeze life. Additionally, well productivity can be an important guide because any scale buildup will be expected to begin to impair the well.

The returning inhibitor concentrations should also be measured so that they can be compared with the MIC determined from laboratory tests. Some preservation might also be needed to avoid these measurements changing over time. In some cases, there could also be difficulties associated with the accuracy of measurement and in the detection of low concentrations (the MIC will inevitably be a relatively low concentration compared with those seen early in squeeze life), although the detection capabilities of many laboratories have improved significantly. Furthermore, as has previously been indicated, it is possible that a retreatment might need to be performed to protect parts of the well even though average inhibitor levels remain greater than the MIC.

There are two primary methods used to quantify the MIC, neither of which is ideal because both are often far removed from the true field situation. The first is a static jar or bottle test in which the concentration of scaling ions is measured at the required conditions and in the presence of various inhibitor concentrations. The second test used is a dynamic tube-blocking test in which incompatible waters are mixed before being passed through a capillary, with the pressure drop across the capillary being measured. Both tests have strengths and weaknesses, and it is thus recommended that both be applied to test candidate inhibitors.

The merit of these tests lies in the fact that the resulting estimated MIC has been found to be in reasonable agreement with that required to control scaling in the field. Therefore, the methodology provides both a useful ranking of inhibitor products and a reasonable indication of the MIC required in the field. It should be stressed, however, that MICs should be optimized in the field. More frequently, laboratory-determined MICs are higher than those observed in the field.

4.3.2 Sulfate Removal. It is evident that there are uncertainties associated with surveillance for resqueeze. Those uncertainties are likely to be magnified in subsea wells, where it could be impossible to collect a water sample from an individual well. Furthermore, treatments are much more difficult and much more expensive to perform in that setting. These problems have led to the development of an alternative scale-control strategy that is primarily used in subsea settings.

This alternative option to control sulfate scaling is to use SRUs to remove the bulk of the sulfate from the injected seawater (Jordan et al. 2006). This equipment is relatively bulky and thus materially increases the footprint associated with the injection plant. Furthermore, because it does not remove all the sulfate ions, it can still leave a residual scaling risk, although the need for squeezing can be expected to be materially reduced, even if it is not removed completely. However, looking at the scaling risks for SRU quality water across a range of barium-ion concentrations, it can sometimes be observed that although very low sulfate concentrations would need to be delivered to fully eliminate the thermodynamic scaling risk, the kinetics of reaction could be such that downhole scale might not be realized and that scale might be manageable simply by means of scale inhibition at wellheads (Simpson et al. 2005). The work of McElhiney et al. (2006) supports that view, suggesting that at high temperatures, the kinetics of reaction at sulfate levels of 40 to 50 mg/L are such that nucleation might take 1 to 3 hours, thereby rendering effective near-wellbore protection even when a thermodynamic scaling risk is present.

It is also worthwhile to consider the levels of sulfate removal needed to eliminate scale squeeze requirements, taking into account in-situ stripping reactions. Mackay et al. (2005b) found a rate law, which was used to define safe operating envelopes based on the barium concentrations of formation waters. These results are summarized in **Fig. 22** for different assumptions of the rate of reaction.

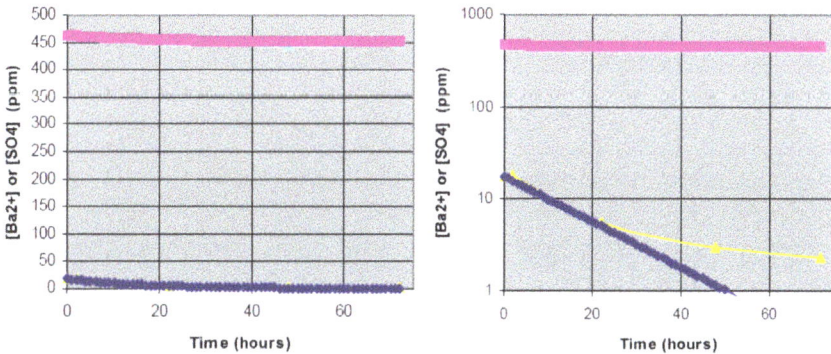

Fig. 22—Safe operating envelopes based on sulfate- and barium-ion concentrations (Mackay et al. 2005b).

Although this analysis might provide a useful guide as to the levels of sulfate needed to protect subsea wells, it is nevertheless appropriate to perform a more robust analysis on a case-by-case basis to ensure that appropriate levels of sulfate removal are applied.

Despite the cost of sulfate removal and the operational difficulty associated with the management of this equipment, the number of SRUs being used in the industry has mushroomed in recent years. SRUs also provide the supplementary benefit of significantly reducing the reservoir-souring risk. Because deepwater producers can cost more than USD 100 million, it is critical to protect such wells, and the number of SRUs in use in subsea fields is therefore a testament to their success in addressing scaling problems in subsea producers.

The SRU equipment uses a nanofiltration process and, as such, requires extremely clean water to avoid fouling of the membranes. Thus, the issues associated with the design and operation of this equipment are by no means straightforward (see *Waterflooding: Facilities and Operations*, another book in this series). Consequently, its incorporation into the process can potentially have negative impacts on the uptime it can deliver. An additional problem is that the membranes are incompatible with the THPS biocide now used in most seawater-injection systems. However, biological control is a critical requirement to avoid biofouling of the filters. Most vendors recommend using 2,2-dibromo-3-nitrilopropionamide as the biocide for this application.

These issues therefore suggest that although sulfate removal offers some clear benefits for scale management in deepwater settings, it is appropriate to balance the risks against other factors such as operability and the economics of the process (Jordan et al. 2001; Jordan and Mackay 2005).

When SRUs are used to control scale (or, indeed, to control reservoir souring), care must be taken to ensure that the mechanical oxygen-removal efficiency is reasonably good to avoid the overuse of the chemical oxygen scavenger. This is because bisulfite-based chemical scavengers convert to sulfate when they scavenge oxygen, and therefore, if excessive dosages are required, they reintroduce the problems that the SRU was designed to avoid. (This is a problem only for a system configuration in which the sulfate removal is located upstream, rather than downstream, of the deoxygenation equipment.)

In the early deployments of this technology, which began in the North Sea Brae Field in 1988 (Heatherly et al. 1994), the intention was not to completely remove the sulfate-scaling risk but to reduce it to more manageable levels. In all cases, desulfation of the injected seawater delays the time of sulfate breakthrough to higher seawater fractions because of scale deposition within the reservoir. **Fig. 23** shows that the maximum scale is shifted to higher seawater fractions.

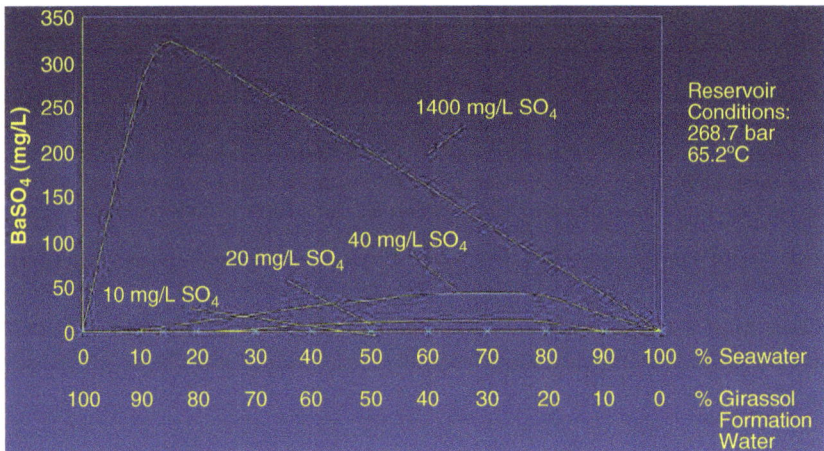

Fig. 23—Quantity and timing of scale vs. sulfate level (Pedenaud et al. 2012).

The use of sulfate-removal membranes as a scale-control technology has been documented for the Girassol Field, offshore Angola (Vu et al. 2000). Barium and strontium levels in the formation water were both 230 mg/L, so the sulfate-scaling risks were appreciable. However, scale squeeze treatments were not considered reliable for this field because of the remote network of daisy-chained wells, high reservoir permeability, and long perforation intervals. In early applications of sulfate-removal technology, there was a reduced but not eliminated scaling risk. In this case, the operator chose to apply a sulfate removal in which levels were reduced to 20 mg/L.

More recently, improvements to the technology have resulted in deployments where the sulfate-removal efficiency is improved and the intention is to completely remove the scaling risk in the producing wells. Pedenaud et al. (2012) provide a summary of the experience of a major operator with four SRUs over more than 10 years of operations. The main driver was a significant barium sulfate and strontium sulfate scaling risk in subsea wells, where problems with the placement of scale squeeze and difficulties with subsequent monitoring to define the optimal resqueeze frequency are key concerns.

The operator had aimed to use the SRUs to eliminate the need to squeeze. Scale prediction studies for the Girassol Field case had predicted that sulfate levels of 10 ppm would be needed to eradicate the scaling risk. The SRU vendor indicated that a two-array sulfate-removal plant could deliver 40 ppm of sulfate, and the decision was made to use this design for the first unit on the basis that much of the sulfate would be removed in the reservoir through precipitation. At that time, an injection line in the producing wells was retained as a backup to limit the risks. Over 10 years, sulfate levels of 20 to 30 ppm were consistently achieved, and the sulfate level in the produced water remained zero, indicating a full stripping of this residual sulfate in the reservoir. Based on this experience, the operator no longer provides a chemical injection line in the producing wells of new developments. Operationally, the experience was also a positive one because a high degree of importance was placed on the need for close operational attention and skilled operators.

Where sulfate removal is used, there will come a time, as the flood matures, when sulfate removal will not be required from a scale-prevention perspective. Such a scenario could allow the sulfate-removal equipment to be retired, which would be beneficial from the perspective of the impact on operating cost; it would also be expected to result in a positive impact on the injection uptime. Conversely, it might have some negative impacts because it could result in an increased souring risk. Depending on the maturity, that might not necessarily result in the production of H_2S, especially if the injectors are located peripherally. It might be appropriate, therefore, to perform reactive transport modeling (both the scaling and souring risks could potentially be assessed) to decide if the retirement of sulfate-removal equipment is possible and to specify the timing at which this would be possible. Al-Riyami et al. (2008) documented these considerations as they applied to the Tiffany Field in the North Sea, and such studies should be undertaken by all operators of this equipment as the field matures, assuming that the equipment has been installed for scaling control rather than souring control.

4.4 Scale Remediation. Even with the very best of intentions, it is not always possible to operate in a completely preventative mode when it comes to scaling. It is therefore inevitable that some degree of scale remediation will be required in many fields. This can occur for a number of reasons, including

- The scaling risks are not properly understood.
- The scale-prevention method cannot be adequately deployed at the optimal location.
- The scaling risks are so severe that it is not possible to fully prevent scaling even if a robust control program is implemented.

For these reasons, many fields need to consider appropriate scale remediation options.

4.4.1 Mechanical Removal. The mechanical removal of scale can be achieved using a range of options, including jets (Johnson et al. 1998), scrapers, brush pigs, or milling operations (Enerstvedt and Boge 2001). Mechanical methods work well for the removal of deposits from the wellbore but do not work well if the near-wellbore region has been subject to significant scale formation.

4.4.2 Dissolution. Because of the limitations associated with mechanical scale removal, chemical methods of scale removal are often preferred. Scale can often be removed by soaking the scale in an appropriate solution. The solution tends to be more effective when the scales are soft because they can be easily penetrated. Carbonates and some sulfide scales can be dissolved by acids, but as previously mentioned, older iron sulfide scales tend to convert to forms that contain more sulfide than iron, and these no longer have good acid solubility. The pH will increase as the scale dissolves, and this will limit the amount of scale that can be removed with each pass of solvent.

When removing sulfide scales with acid, it must be remembered that the acid treatment will convert the sulfide back to H_2S, which is likely to be back produced after the treatment. There are therefore clear HSE issues to be managed in such treatments. In these cases, it might be necessary to include an acid-insensitive H_2S scavenger into the formulation.

For sulfate scales, a treatment with a chelating agent is commonly required, and the scale is soaked with a solution containing the chelant. In such treatments, the chelant forms a complex with divalent cations from the scale. This shifts the equilibrium between scale and chelant, and thus, more scale continues to dissolve.

Commercial scale dissolvers are often based on amino-carboxylic acids such as ethylenediaminetetraacetic acid (EDTA) (De Vries and Arnaud 1993) or diethylenetriaminepentaacetic acid (DTPA) (Putnis et al. 1995), which have been shown to effect the successful dissolution of barium sulfate when applied appropriately. As well as removing wellbore scale, dissolver treatments can also be expected to remove any scale deposited in the near-wellbore region. Consequently, such treatments might also feature a stimulation of well performance (Lejon et al. 1995). However, these highly alkaline, surface-active chelation packages could potentially cause mineral dissolution and formation damage, especially in cases where clays are present (Jordan et al. 1998).

The factors that influence the effectiveness of these treatments include the dissolution rate, temperature, placement, impact of oil, treatment concentration, and pH control.

Dissolution Rate. Scale removal occurs in two steps. First, dissolution of the scale in the scale dissolver occurs, in a slow reaction for scales such as barium sulfate that have low-solubility products. The chemical reaction between components in the scale dissolver and the dissolved-scale ions is then a fast one that consumes the dissolved ions and therefore helps to drive the equilibrium in the first step.

Temperature. The rate of dissolution of barium sulfate scales using 0.05 and 0.5 molar DTPA solutions at different temperatures has been studied (Putnis et al. 1995). The dissolution rate of barium sulfate is highly temperature dependent, with a reported activation energy of -45 $kJ \cdot mol^{-1}$. This value suggests that the rate is controlled by the desorption of a barium-DTPA surface complex. This is further supported by the observation that the initial dissolution rate is inversely related to the DTPA concentration, and the 0.05 molar DTPA solution was found to be more efficient as a solvent than the 0.5 molar solution. This was interpreted as a result of passivation of the barium sulfate surface by the formation of a surface complex layer at high DTPA concentrations. These results are shown in **Figs. 24 and 25.** At temperatures below 70°C, the rate of removal for sulfate scales is usually found to be too slow for such treatments to be practical.

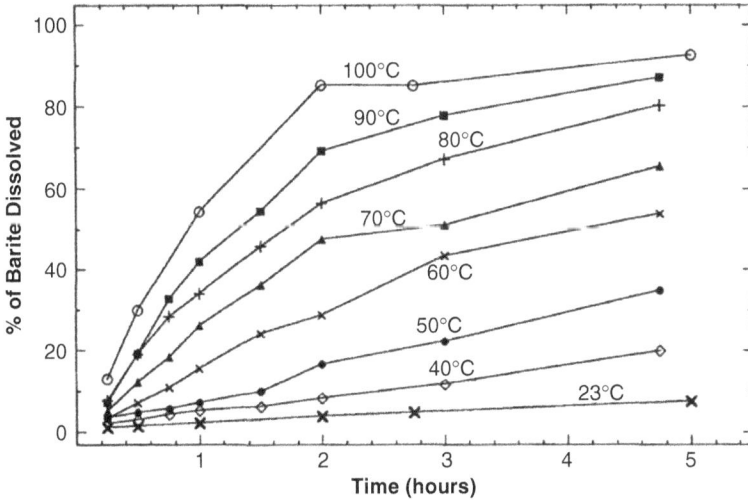

Fig. 24—Dissolution as a function of temperature in 0.5 molar DTPA solution (after Putnis et al. 1995).

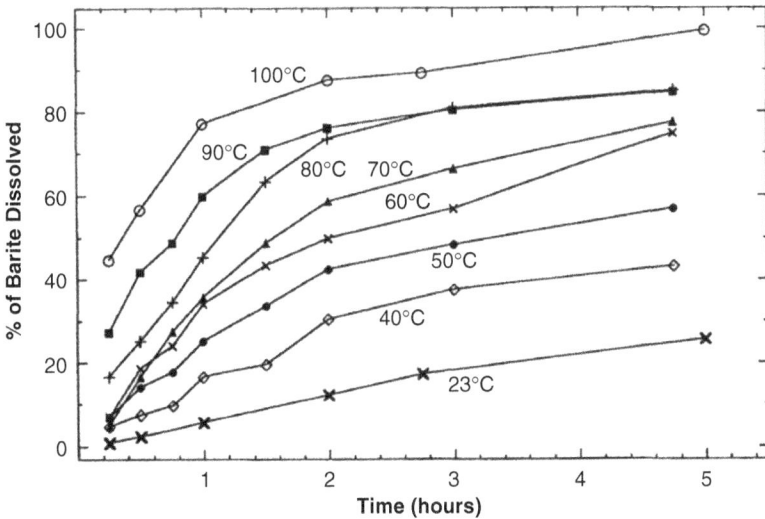

Fig. 25—Dissolution as a function of temperature in 0.05 molar DTPA solution (after Putnis et al. 1995).

Placement. Even when temperatures are favorable, a contact soak time of 12 to 48 hours is typically required to effect dissolution. Furthermore, treatments tend to be more effective in cases where the dissolving solution is kept in motion. In practical terms, the use of chelating agents for the treatment of sulfate scales is best suited to the treatment of small scale volumes, so the removal of scale from a safety valve, for example, would be a suitable application.

Impact of Oil. When an oil film covers the scale, contact between the scale and the scale dissolver might be inhibited. As a result, the reaction rates could be appreciably lower than expected. Many treatments therefore use a preflush step before performing the scale-dissolver treatment. Such stages usually consist of a mutual solvent and might also incorporate a surfactant.

Treatment Concentration. Most barium-scale dissolvers are found to be more effective when diluted concentrations are used.

pH Control. The pH should be kept high throughout the treatment (greater than the pKa of the chemical) for optimal performance. For most chemicals, this means a pH greater than 9 to 10 (often achieved through the use of potassium hydroxide).

Because a residual scale-deposition risk is frequently still present at the time dissolver treatments are performed, scale-dissolver treatments are often combined with scale squeeze treatments (Smith et al. 2000).

Because iron sulfide scales are very often coated in oil, treatments for this scale often require the program to incorporate a phase to remove the organic component to enable the treatment to reach the iron sulfide scale itself.

4.5 Scale Types and Characteristics.

4.5.1 Calcium Carbonate.
Calcium carbonate is the most commonly seen scale because many formation waters are saturated with this mineral. It is therefore found in both sandstone and carbonate reservoirs, and it might be an appropriate starting point to assume at the outset that the formation water is saturated with calcium carbonate.

The changes in a production process whereby both temperature and pressure are observed to drop are conflicting in terms of the effects on calcium carbonate solubility because, although its solubility decreases as the pressure drops, calcium carbonate unusually exhibits increased solubility at lower temperatures. Most minerals have increased solubility at high temperatures, but calcium carbonate has an inverse relationship because of the impact of CO_2. This gas has lower solubility in water at higher temperatures, and as it is lost, there is an impact on the equilibrium between carbonate (CO_3^{2-}), bicarbonate (HCO_3^-), and CO_2, which plays a significant role in the calcium carbonate deposition process. Thus, deposition resulting from temperature change can be encountered as a result of localized heating, such as at the motor where ESP pumping is used.

Scaling usually occurs as a result of an increase in pH caused by the loss of the CO_2 from the water into the gas phase, which implies that in producing wells, calcium carbonate deposition always occurs downstream of the bubblepoint. As pressure drops below the bubblepoint, CO_2 is lost into the gas phase. That has the effect of driving Eq. 14 farther to the right-hand side, and as a result, the solubility of calcium carbonate decreases.

$$Ca^{2+} + 2HCO_3^- \leftrightarrow CaCO_3 \text{ (s)} + H_2O + CO_2. \dots\dots\dots\dots\dots\dots\dots\dots\dots(14)$$

In the past, scaling calculations tended to only consider reactions occurring in the water phase. However, the flashing of gases from both the oil and the brine phases is critical to a more robust understanding of calcium carbonate scale formation. In addition, it is also important to consider the partitioning of CO_2 gas between the oil and the brine phases during the gas-flash process (Vetter and Farone 1987). These issues must be factored in if a robust scale prediction is to be achieved.

Additionally, other minerals present in the water can have an important effect on scale deposition. Østvold and Randhol (2001) found that both magnesium and sulfate ions resulted in an increase in the time taken for scale to form. This is because both ions tend to be incorporated into the calcite lattice, thereby increasing calcite solubility, and as a result, the less problematic aragonite mineral becomes the kinetically favored scale precipitate.

Because many produced waters are saturated with calcium carbonate, there can be self-scaling risks when this water is used for injection purposes. Although the water might be saturated, most projects suffer no problems under dynamic operating conditions. When water injection is suspended, however, there could be serious scaling problems that, in extreme cases, might cause the plugging of injection lines. Furthermore, scaling problems can sometimes occur as the injection water enters the formation. For one case in Mexico, a pilot project resulted in an injectivity reduction from 4,000 to 1,920 BWPD over a 167-day period (Pruess et al. 2006). Modeling suggested that the fast deposition of calcite in the near-wellbore region of the injector limited the damage zone to a zone within 2.5 m of the well; assuming an initial permeability of 50 md, it was estimated that, despite only a modest porosity reduction, the effective permeability would be reduced to less than 1 md within 130 days.

These observations suggest that continuous inhibition of the injection water would be an appropriate precaution for such conditions.

The most fortunate aspect of calcium carbonate scaling is that it is readily remediated because the scale is easily dissolved in mild acids. The scale can sometimes be oil-coated, so it might be necessary to incorporate a solvent or surfactant in the acid-treatment design to ensure effective contact between the scale and the acid. Hydrofluoric acid should never be used for the removal of calcium carbonate scales because it can result in the formation of insoluble calcium fluoride.

Calcium carbonate has several different potential morphologies. Calcite is the most stable crystalline form, but it does not exhibit the fastest precipitation rate; therefore, kinetically, vaterite (disk shaped) and aragonite (needle shaped) can sometimes form first, especially at higher temperatures, although over time they convert into the more stable calcite (cubic) structure.

4.5.2 Calcium Sulfate. This type of scaling is commonly found in seawater floods where sulfate from the seawater mixes with the high calcium concentrations found in the formation water. The temperature dependence of sulfate scales is generally less important than the impacts of pressure; temperature changes usually only induce relatively mild supersaturations. Because the water solubility of calcium sulfate is appreciably higher than that of both strontium sulfate and barium sulfate, it forms the largest amounts of solid when it becomes supersaturated. Consequently, it is one of the few scales capable of inducing scaling rates so high that production from a well can be killed within a matter of days. Calcium sulfate is a softer scale than either barium sulfate or strontium sulfate, so remediation by means of jetting can be effective; however, it is insoluble in common acids, so chemical removal can only be effected with chelating agents. Only a small amount of supersaturation is needed to induce scaling-deposition problems with calcium sulfate.

4.5.3 Barium Sulfate and Strontium Sulfate.

4.5.3 Barium Sulfate and Strontium Sulfate. These scales form in a manner similar to that of calcium sulfate scales. Calcium, strontium, and barium are in the same periodic group of elements, but the solubility of these elements decreases in that order. This means that barium sulfate has a very low solubility and can still form even with very low barium concentrations in the formation water. The reduction in solubility moving down the periodic group occurs because the size of the ions is increasing. The lattice energy of barium sulfate, in particular, is more than the enthalpy of solvation resulting from the large size of both the anion and cation; therefore, barium sulfate has a very low solubility—0.197 g/L at 20°C in pure water. Solubility improves slightly in the presence of brines (**Fig. 26**) (Yuan 2001), but because formation water might contain hundreds of milligrams per liter of barium ions, it is easy for this scale to precipitate in the presence of seawater that contains sulfate ions.

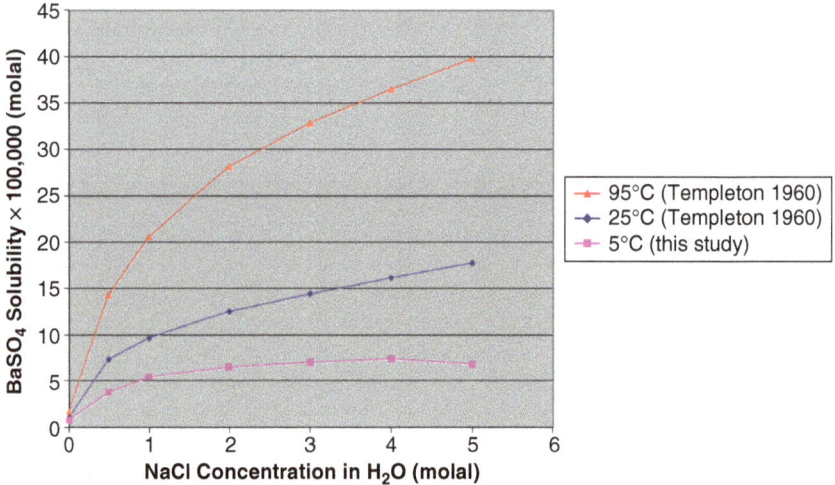

Fig. 26—Barium sulfate solubility in the presence of sodium chloride (after Yuan 2001).

Furthermore, these sulfate scales can coprecipitate with radium sulfate, which is radioactive, introducing an additional complication.

The barium and strontium sulfate scales tend to be hard and can be very difficult to remove mechanically, so prevention assumes a greater importance. It is possible to remove these scales using sulfate-scale solvents, but dissolution rates are slow at lower temperatures, so temperature is a key criterion to decide if a solvent program could be successful. Coiled-tubing-conveyed jetting can be effective in tubulars, particularly if the scale is not pure barium sulfate. Otherwise, if the downhole temperatures are too low to consider a solvent treatment, milling and reaming is the only practical solution (Brown et al. 1991).

4.5.4 Naturally Occurring Radioactive Materials. Radium (Ra) is in the same periodic group of elements as calcium, strontium, and barium. This element has no stable isotopes, but there are 33 known isotopes ranging from ^{202}Ra to ^{234}Ra. The longest-lived and most common isotope of radium is ^{226}Ra, with a half-life of 1,600 years, which is formed in the decay chain of the uranium (U) isotope ^{238}U. ^{228}Ra can be formed from the decay of the thorium (Th) isotope ^{232}Th. Because of its similarity to strontium and barium, radium tends to coprecipitate in the scales that these elements form. Consequently, barium and strontium sulfate scales are sometimes found to be radioactive; Krishnan et al. (1994) describe a case at the Countess "O" Field in southern Alberta, Canada, and Shuler et al. (1995) note another case at Eugene Island in the US. This latter example suggests that a focus on scale prevention is a key to scale management when radioactive scales are encountered. This could well be true because of the difficulties associated with the management and disposal of such scales.

4.5.5 Iron Sulfide. Iron sulfide is often found mixed with bacterial biomass in injection wells, where it forms as a result of iron-containing corrosion products reacting with H_2S generated by reservoir souring. This suggests that perhaps the best means to control the risk of iron sulfide scales is to prevent the reservoir-souring process from occurring in the first place. When iron sulfide scale is encountered, it is important to ensure that the scale is not allowed to dry out when it is removed because the scale is pyrophoric (meaning it can self-ignite), so keeping it wet is important to help minimize the dangers of fire.

The problems associated with iron sulfide can be compounded by the fact that it tends to initiate as very small dispersed particles. Thus, the presence of this scale—sometimes found in produced-water injection systems—is often characterized by an opaque, blackish tinge to the water, which can intensify as the water is allowed to stand. The surface of the sulfide is commonly hydrophobic in the presence of hydrocarbons and thus the scale tends to absorb organic particles. This makes the particles rather sticky, and they then tend to agglomerate. Such agglomeration often occurs after the solids have moved through the water-treatment process, and this could induce injectivity problems even in cases where rigorous solids-removal processes are used. Iron sulfide accumulation combined with hydrocarbons is known as schmoo. This phenomenon can be particularly problematic and is discussed in the next section.

Iron sulfide can be present as a range of different chemical types because both the iron and the sulfur have more than one oxidation state. Therefore, its exact composition will be dependent on temperature, pressure, pH, and H_2S concentration. Compounds that can be formed include, but are not limited to, mackinawite [$FeS_{(1-x)}$], cubic ferrous sulfide (FeS), troilite (FeS), pyrrhotite [$Fe_{(1-x)}S$], greigite (Fe_3S_4), and pyrite (FeS_2).

The age of the deposit can impact the type of iron sulfide present because the continuous exposure of iron sulfide compounds to increasing amounts of H_2S will result in the formation of compounds that have more sulfide than iron in the structure. The type of iron sulfide encountered is important because each type has different properties; mackinawite and troilite have good solubility, whereas pyrite is very insoluble. Standard hydrochloric acid will easily remove iron sulfide species in which the iron and sulfur are present in molar ratios close to 1, but older iron sulfide scales, in which there is more sulfide than iron (e.g., FeS_2), are insoluble in acids (Nasr-El-Din and Al-Humaidan 2001).

In cases where species of sulfur-rich compounds—such as pyrite—are encountered, a mechanical treatment is commonly used as a precursor to an acid treatment. Acid treatments also typically use a mutual solvent because most of these deposits contain some amounts of hydrocarbons.

Iron sulfide control could involve a combination of removal/dissolution and prevention strategies (Wylde 2014).

Iron sulfide can be removed mechanically using coiled tubing (Bittner et al. 2000). Mechanical pigging could be a potentially viable methodology for removal, but injection systems rarely have such capabilities. Consequently, removal has largely been achieved using chemical treatments. Chemical removal can be achieved using hydrochloric acid (except for sulfide scales with more sulfide than iron), although it has drawbacks because the rapid generation of a significant amount of H_2S means that a high concentration is then released into the process, in addition to the potential for corrosion, and this suggests chelating-type treatments might be preferable (Yap et al. 2010).

A surfactant could also be needed to remove suspended oil and enable the acid to reach the scale. The challenges associated with acid treatments have led to the emergence of chelating agents as effective removal agents. These include DTPA, THPS, and EDTA (Yap et al. 2010). Acrolein can be a very effective iron sulfide dissolver (Salma 2000) as well as a very effective H_2S removal agent, but there can be significant handling issues with this chemical.

The biocide THPS, which is the primary biocide used in many water-injection systems, has also been reported to be an effective dissolver of iron sulfide scales (Larsen et al. 2000; Talbot et al. 2002). This suggests that THPS might be the best choice for an injection biocide because it could also help to control biological populations.

Prevention strategies can use a blend of different techniques, including

- Chelating agents for iron sequestration
- Surfactants for water-wetting
- Biocides to target biofilm
- Corrosion inhibitors to lower the total amount of iron in the system
- Threshold scale inhibitors

4.5.6 Schmoo. The previous section noted that where hydrocarbons or other organics are present, the iron sulfide could become coated in those organic materials and agglomerate to form schmoo. It is interesting that schmoo is not always encountered in PWRI schemes where H_2S is present. The distribution of total sulfide between the dissolved H_2S gas and the ionized form depends on temperature and pH. At low pH levels, most of the total dissolved sulfide exists as H_2S molecules, whereas at elevated pH levels, most of the total dissolved sulfide exists as ionized sulfide, primarily HS⁻, with S²⁻ being found only at a very high pH. Assuming that the water is saturated with iron and sulfide ions, as the pH rises, the solubility of iron sulfide decreases. This could induce small particles to come out of solution that are then rapidly contacted by oil droplets to create schmoo. This raises the possibility that it might be feasible to control schmoo using noninterfering pH buffers (Przybylinski 2001). In a situation analogous to calcite scaling, keeping the gas phase to a minimum might also be expected to have beneficial impacts.

The critical pH value might be approximately 6.5—with iron sulfide unlikely to form at a pH less than this value but likely to form at a pH greater than this value. Wylde (2014) also notes that production chemicals that have a significant influence on the pH, such as H_2S scavengers, could therefore also have an impact on the likelihood of iron sulfide formation.

Wylde (2014) adds an additional relevant factor to the discussion. The redox potential—a measure of the potential of the solution to gain or lose electrons—could also be important. The lower (i.e., more negative) the redox potential, the more reduced the system is, which makes it more likely to be oxidized and gain electrons. The redox potential must be low for reduced sulfur species to be present, which might then contribute to sulfide-scale formation.

Acrolein is another option to treat schmoo problems in cases where dissolution of the iron sulfide is needed (Salma 2000). This chemical has a number of beneficial features: It not only dissolves iron sulfide, but it also acts as an H_2S scavenger and is a very effective biocide. However, there is a drawback because there are HSE concerns associated with its use. (It is an inhalant poison.) Nevertheless, it is somewhat surprising that this option has not been used more widely.

A 9-week trial of acrolein treatments was conducted in a water-injection system in Oman (Horaska et al. 2009). Based on biocidal effects, a 90% reduction in planktonic SRB and a 72% reduction in planktonic general aerobic bacteria were reported. The dissolved H_2S concentrations were reduced by 50%, inducing a downward trend in oil free suspended solids. Furthermore, a significant improvement in the injection-water quality resulted in a 16% increase in injection volumes.

At Prudhoe Bay in Alaska, schmoo problems prompted the development of a dispersant to remove the schmoo (Ly et al. 1998). The dispersant consisted of two nonionic surfactants—an alkyl polyglycoside and a linear alkyl ethoxylate—dissolved in an aqueous solution of sodium hydroxide.

5. Incompatibility with Clay Mineralogy

Clay minerals are a common feature in sandstone reservoirs. An understanding of their structure is key to understanding how they respond as injection water is introduced to the reservoir. They are a group of hydrous aluminum phyllosilicates, but a number of different elements can be incorporated into the structure, which invariably comprises 2D sheets. The most important clay minerals in a waterflood context are smectites (including montmorillonite), illite, kaolinite, chlorite, and mixed-layer clay minerals. The clay structures consist of both tetrahedral and octahedral sheets. The tetrahedral sheets consist of four oxygen atoms at the apex locations of a regular tetrahedron, connected to a silicon atom at the center. An interlocking array of these structures connected at three corners in the same plane by shared oxygen atoms forms a tetrahedral sheet network. Some of the tetrahedrons could have an aluminum atom at the center instead of a silicon atom, and occasionally they could have iron, or another element, instead.

The octahedral structure features hydroxyl (OH⁻) ions or oxygen at the corners of the octahedron and a cation at the center, which is usually aluminum, iron, or magnesium. The octahedral sheet is formed by sharing all hydroxyl groups at the corners of an octahedron with neighboring octahedrons.

The way that the tetrahedral and octahedral sheets combine determines the clay type. In some cases, a single tetrahedral layer combines with a single octahedral layer. For this case, if the apical oxygen of the tetrahedral sheet replaces one hydroxyl of the octahedral sheet, this is the structure of the clay type kaolinite, which has the general formula $Al_2Si_2O_5(OH)_4$.

When an octahedral sheet is sandwiched between two tetrahedral sheets (a 2:1 clay), a three-sheet mineral type results, with two-thirds of hydroxyls in the octahedral sheet between two tetrahedral sheets replaced by apical oxygens of the tetrahedral sheet. When water molecules and cations occupy the space between the layers, a smectite/montmorillonite clay is created. However, the size and charge of potassium is such that it fits neatly in the hexagonal ring of oxygens of the adjacent silica tetrahedral sheets, and thus, when potassium is present, the structure features a strong interlocking ionic bond that holds the individual layers together and prevents water molecules from occupying the interlayer position as they do in smectite. This then is the clay type illite, which in simple terms could be described as potassium smectite.

Chlorites also have a 2:1 sandwich structure, but in their case, the space between the sandwich structures is filled by $(Mg^{2+}, Fe^{3+}) (OH)_6$, which is often called the brucite layer; this is sometimes said to represent a 2:1:1 structural arrangement.

Mixed-layer clays can be any combination of clay types in which the individual clay crystals are stacked on top of one another in an irregular arrangement; illite-montmorillonite is the most common mixed-layer clay encountered.

The metal atoms in the clay lattice can be substituted. If an atom of aluminum (Al^{3+}) is replaced by an atom of magnesium (Mg^{2+}), a charge deficiency of 1 results, and this charge is compensated for by cations located in the interlayer region, which can be freely exchanged. The cation-exchange capacity of the mineral depends on crystal size, pH, and the type of cations involved. The following presents the order of replaceability of the common cations in clays:

$$Li^+ > Na^+ > K^+ > Rb^+ > Cs^+ > Ca^{2+} > Sr^{2+} > Ba^{2+}$$

Thus, sodium montmorillonite swells more than calcium montmorillonite because the calcium cation is more strongly adsorbed compared with the sodium cations.

The adsorption of water molecules on the basal crystal surfaces (on the external and, in the case of expanding minerals, the interlayer surfaces) occurs by means of hydrogen bonding, holding a layer of water molecules to the oxygen atoms, which are exposed on the crystal surfaces. Thus, water molecules surround a clay crystal structure and position themselves in such a way as to increase the spacing between layers, resulting in an increase in the structure's volume.

Table 3 demonstrates that the internal surface area of smectite is much greater than that of other minerals; hence, this clay type is much more prone to swelling.

Table 3—Key properties of different clay types.

Mineral Type	Surface Area (m²/g)			Cation-Exchange Capacity (meq/100 g)
	Internal	External	Total	
Smectite	750	50	800	80–150
Illite	5	15	30	10–40
Kaolinite	0	15	15	1–10
Chlorite	0	15	15	< 10

Two types of clay problems are generally encountered. In the first type of problem, primarily in illite and kaolinite, the clays dislodge from the sand grains and the fines migrate farther (either deeper into the reservoir in an injector or toward the perforations in a producer), where they can then plug pore throats. The second problem is associated with montmorillonite and results in an irreversible expansion of the clay.

It has been noted that the basic montmorillonite building unit consists of an octahedrally structured aluminum sheet sandwiched between two tetrahedral silicate sheets. These layers have a negative surface charge, and the layers are stacked together with an interlayer space containing hydrated charge-compensating cations. It is the negatively charged clay surfaces that are responsible for the sensitivity to injection water. Sodium, potassium, or calcium ions in the connate water attach to the clay surface by means of electrical attraction, effectively neutralizing the negative charges on the surface of the clay. In this state, the clay is stable. The introduction of a less saline water could dilute the connate water. As the cation cloud surrounding the clay becomes more diffuse, the water molecules can enter between the clay platelets, resulting in a swelling that can be irreversible in a waterflood context (**Fig. 27**). In the most extreme case, up to 18 layers of water molecules are able to enter the structure, which can result in a swelling of up to 600%.

For the clay types susceptible to fines migration, the mechanism is essentially the same, but the effect is slightly different. As is the case for montmorillonite, the clays have a net negative charge, and in an electrolyte solution, these clays are surrounded by positive cations (the counterion cloud). The distribution of this cloud is determined by the opposing forces of electrostatic attraction to the negative clay lattice and diffusion into the surrounding bulk solution. In concentrated electrolyte solutions, the counterion clouds are small, allowing the clay particles to closely approach one another. The clays are kept flocculated by van der Waals or dispersion forces. In dilute brines, diffusion causes the counterion cloud to expand. Hence, interparticle distances are large and the dispersion forces are not effective, and as a result, the clay particles disperse (**Fig. 28**).

It is possible that a formation-water salinity reduction is not the only mechanism that can induce fines mobilization. There are suggestions that reductions in oil saturation as the waterflood front passes could potentially initiate clay-related problems, possibly because some clays might have been contacted by oil and then, because of the saturation change, become open to contact with water. Muecke (1979) observed that for fines that are water-wet, any movement of that wetting phase can induce

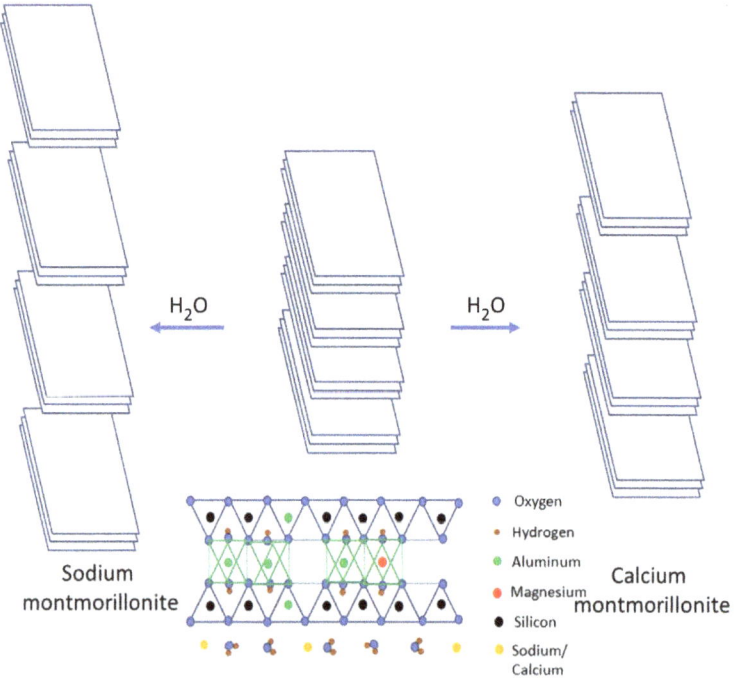

Fig. 27—Clay-swelling mechanism for montmorillonite.

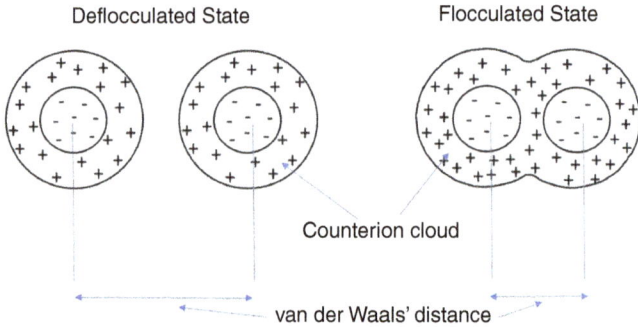

Fig. 28—Mechanism of fines migration.

fines migration. Consequently, any movement of connate water, which will occur as a bank ahead of injected water, could still induce fines migration and, hence, might potentially impair producers.

To assess the risks of clay-related problems in a waterflood, the first step is to assess the types of clay present in the reservoir and their distributions. This is commonly conducted by performing X-ray diffraction analysis on core material.

The impact on clays that might be expected as injection water starts to replace formation water can be assessed on the basis of work conducted in the 1980s. Scheuerman and Bergersen (1990) identified two criteria that need to be met to prevent the deflocculation of formation clays and the subsequent associated formation damage:

- The water must have adequate total cation and/or divalent-cation concentrations.
- Cation exchange in the transition from one water to another must not reduce the divalent-cation concentration below the level required to keep the clays flocculated.

Based on this work, Scheuerman diagrams were produced to identify the compositional changes that might be dangerous for each of the different clay types (**Fig. 29**). The diagrams effectively defined the conditions under which problems might be encountered on the basis of an analysis of the divalent-cation concentration as a function of the total cation concentration. For an injection water to be ensured of compatibility with formation clays, its cation must be such that it falls into Zone A of the flocculation salinity diagram for the most water-sensitive clay found in the reservoir. The divalent-cation concentration is important because the clay-flocculating power is a function of valence (charge) and increasing ionic radii. Generally speaking, divalent cations are found to be 50 to 100 times more effective than monovalent ions at flocculating clays. An increasing calcium-ion concentration thus sharply reduces the flocculation salinity.

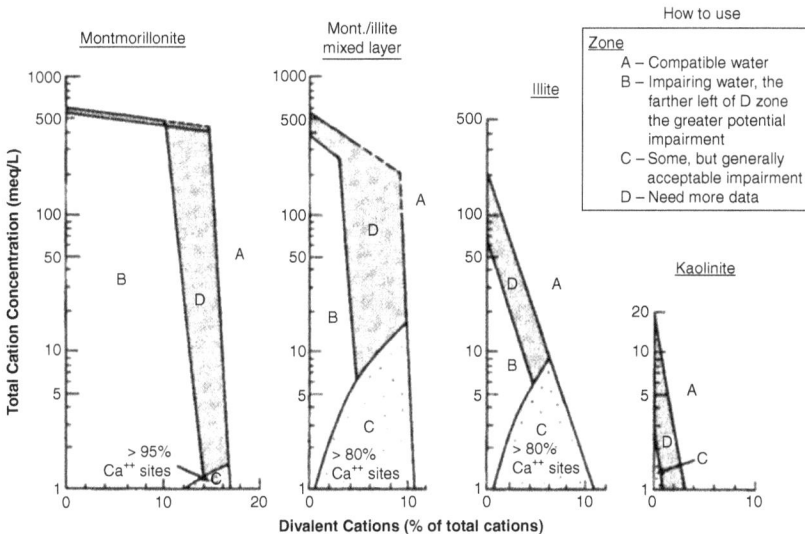

Fig. 29—Flocculation-salinity criteria for various clays (after Scheuerman and Bergersen 1990).

The farther to the left of the flocculation salinity and the lower the total cation concentration, the greater the risk that there will be an impairment problem. Scheuerman and Bergersen (1990) also suggested that for reservoirs with temperatures greater than 121°C, reservoir clays will have little, if any, sensitivity to freshwater injection. It is important to recognize that these flocculation salinities were determined by flowing increasingly diluted brine through a sand/clay pack or core until the clays were deflocculated, as indicated by an injection-pressure increase as a result of clay-particle release and subsequent pore-throat plugging. As such, these experiments do not fully replicate the conditions that would be experienced in a waterflood. Specifically, no oil was present in the core plugs at the time of the tests, and oil might help to somewhat mitigate the dangers of clay swelling or mobilization. As such, these diagrams possibly represent worst-case conditions for the manifestation of clay problems and, in reality, more severe conditions might be needed to induce problems.

It is important to note that Scheuerman and Bergersen (1990) address the risk of clay problems resulting from chemical effects. It is also possible that fines mobilization can be induced as a result of velocity effects. These risks can be assessed by means of core-flush tests.

Many early clay-stabilization treatments were controlled using inorganic compounds containing monovalent cations such as potassium hydroxide (KOH) (Norman and Smith 2000) as a clay-stabilizing chemical. This technique was found to have a short-lived effect, however, because ongoing ion-exchange reactions tend to reduce the protective capabilities of the treatments over time, which led to a reinitiation of the formation-damage mechanisms. Polyvalent salts such as aluminum (Reed 1972) and zirconium (Peters and Stout 1977) were then introduced, and they were found to offer a longer-lasting treatment. However, these results were based on the use of the hydroxide salts, which have very low water solubility. Furthermore, the cations were found to lose their affinity for the clays at high pH levels.

This protection philosophy therefore has gradually been replaced. Current clay-stabilization treatments tend to use cationic polymers (Weaver et al. 2011), with the polymer adsorbing onto the surface of the clay, thereby blocking the subsequent adsorption of water. It should be noted that where montmorillonite swelling is predicted, treatments need to be performed pre-emptively because any swelling could be irreversible. Zhou et al. (1995) suggested that cationic organic polymers are the most commonly used type of chemistry for such treatments, although high-molecular-weight formulations could induce formation damage and there could be limitations for their use in higher-temperature applications. A cationic inorganic aluminum/zirconium-based polymer has been developed that presumably is designed to incorporate the beneficial features of both the polymer and the polyvalent cations (Clarke and Nasr-El-Din 2015).

Although clay-related problems are a recognized issue (commonly found in a number of early US floods, which explains why so much focus was placed on this issue at that time), it must be said that this problem does not seem to be commonly encountered in waterfloods today. Given the mechanism whereby problems occur, however, this should perhaps not be a surprise. Generally, low water salinities are required before any problems manifest. Even when low-salinity water is injected, problems might not always occur, presumably because either the problematic clays are not present or the oil has helped to stabilize the clay. In Syria, a large number of fields were waterflooded, with the water taken from the Euphrates River—an extremely fresh water source—without any clay-related problems experienced. Thus, although it would be prudent to assess the risks of such problems for new assets, the likelihood of such problems being encountered is low for most fields.

5.1 Shale Stability. Shale is a fine-grained sedimentary rock in which clays are a significant component. On the basis of what has been stated in the previous section on clays, it should come as no surprise that the water sensitivity of such materials is a major topic. Shales can be important source rocks for hydrocarbons, but because of their low permeability, they were not viewed as producing reservoirs until recent years, when fracturing technologies began to unlock such resources.

There is still some debate regarding the mechanisms by which water induces shale instability, but based on the mechanisms of clay interactions with water, it might be no surprise to hear that one of the important causes of shale instability is volume expansion following the swelling of smectite. The properties of shales are known to induce significant wellbore-stability problems when wells are drilled with water-based drilling fluids. In recent years, shale stability has also become a key issue in the hydraulic fracturing of shale oil and gas deposits.

Shales have permeabilities that are commonly measured in the nanodarcy range, and stabilization measures taken during drilling are inevitably surface-based. Fortunately, from a drilling perspective, the interval needs to be kept open only long enough to drill the section in question, and then the shale interval is isolated behind cemented casing.

If water-sensitive shales are present in a waterflood context, the problems might not be so easy to solve. Luckily, when sand/shale sequences are encountered, the shales are often found to be quite stable, but this will not always be the case. When water-sensitive clays are found dispersed within the sand in an injection well, a polymeric stabilization treatment can be pumped into the near-wellbore region to protect the well. When water-sensitive shales are present, however, there would be essentially no penetration associated with such a treatment, and because very significant volumes of water will be injected into the well over time, it seems implausible that a surface treatment of the shale would be expected to remain in place to protect the well.

In assessing a waterflood where shales are present within the completion interval, it therefore seems reasonable that a suitable first step would be to assess the sensitivity of those shales to the water that will be injected (see *Waterflooding: Injection Regime and Injection Wells*, another book in this series). To ensure representative results, those tests should be conducted under confining conditions. For cases where a problem is then predicted, the shales can be isolated by casing, cementing, and perforating, provided that the sands and shales are not so thinly bedded that isolation is not effectively possible. However, in cases where sand consolidation is required, the problem is more difficult because the completion normally features an open annulus that precludes isolation.

5.2 Low-Salinity Flooding. The different water sources available for use in a waterflood can have a significant impact on project success. For example, as has previously been noted, reservoir souring is almost inevitable for seawater floods unless preventative measures are taken. Under certain circumstances, the chemical composition of the water can also have a significant impact on the reservoir flood. In recent years, there has been much discussion in the literature in relation to the potential of low-salinity water to increase recovery. There appear to be low-salinity effects seen in both sandstone and carbonate reservoirs, although the mechanisms are different for these two main reservoir types. These issues will be discussed in more detail, but it appears that the understanding of the effects is more mature for sandstones.

It is now well-established that wettability has a significant impact on flow parameters such as relative permeability and capillary pressure and that the wettability is dependent on crude-oil/brine/rock interactions. Early work showed that those interactions and oil recovery depend on brine composition, but the mechanisms by which oil recovery was changed were not well-understood. Interest in low-salinity effects began to increase starting in the late 1990s when work conducted at the University of Wyoming on Berea Sandstone core material suggested that wettability changes could be involved in the mechanism (Tang and Morrow 1997, 1999a, 1999b). Additional work from the same group suggested that low-salinity water causes a reduction in the water relative permeability curve, resulting in an improvement in microscopic sweep efficiency (Maas et al. 2001). The reduction of water permeability, with the observation of fines production from the core, suggested that clays were playing a role in the effect. The role of clays, although not yet confirmed, was also suggested because it had been known for a long time that clays in the reservoir react differently toward waters of different salinities.

Empirical evidence suggests that the low-salinity flood effect is associated with the removal of divalent cations such as magnesium (Mg^{2+}) and calcium (Ca^{2+}) from the water. These cations tend to adsorb to sand grains and particularly to clay surfaces in the reservoir. By adsorbing to these surfaces, the cations tend to make the rock more oil-wet because polar components in the crude oil, such as carboxylic-acid groups, tend to be bound to the rock by means of an ion-bridging mechanism. Consequently, flooding with low-salinity water removes these divalent ions, thereby making the rock more water-wet and releasing surface-bound oil components, which improves the recovery process.

A change of rock wettability into a more water-wet state is expected to have the following consequences:

- It will cause a shift in capillary pressure curves to higher water saturations.
- It leads to an increase in oil relative permeability and a reduction in water relative permeability (Owens and Archer 1971).

Hence, this change leads to an improvement in microscopic sweep efficiency.

Most minerals, including quartz, are generally assumed to be water-wet when clean. However, reservoirs comprise many mineral types, and if the mineral surfaces are contaminated, they could become less water-wet, depending on the chemical character of the contaminant. Clays are important in this respect because their large surface area and high charge density make them among the most reactive of minerals. This means there is a strong likelihood that molecular interactions will occur between the petroleum components and the clays within a petroleum reservoir, and oil components could adsorb to the clays by means of ion-exchange mechanisms, making the rock more oil-wet.

Additional clues to the mechanism are evident in the fact that although a low-salinity effect is observed in most core experiments, it is not observed in all. Furthermore, effects are observed only if the core contains connate water, if the oil phase is a crude oil, and provided that the core has not been "fired" to remove clays (Tang and Morrow 1999b). On balance, the improvement in oil recovery appears to occur by means of changing the wettability toward a more water-wet state. Lowering the overall concentration of cations, and particularly lowering the concentration of divalent cations in the brine, reduces the electrical bridging effect and thus increases the repulsive forces between the clay particles and the oil. Eventually, the oil particles will be desorbed from the surface of the clay, and there will be a change in wettability toward a water-wet state (**Fig. 30**).

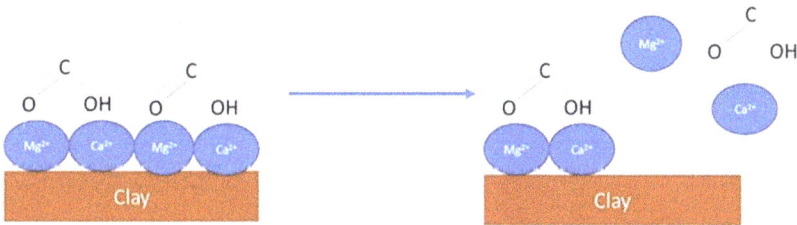

Fig. 30—Likely low-salinity mechanism.

A closer investigation of the wettability alteration by heavy-end adsorption on clay minerals shows that divalent cations will cause more heavy-end adsorption, and an increase is observed in the following order: $K^+ < Na^+ < Ca^{2+} < Mg^{2+}$ (Collins and Melrose 1983). This suggests that wettability is related to water composition. If fresh water is then injected, the existing chemical equilibrium in the reservoir could be disturbed by cation-exchange processes between injected water, formation water, and the clays.

There is a problem here because it has already been noted that low-salinity injection can cause serious problems in relation to injectivity in waterfloods. In experimental studies, it was found that at salinity levels slightly greater than those able to induce formation damage, the injection of fresh water resulted in cation-exchange processes between brines and clays that removed adsorbed polar compounds from the clay surface. However, in contrast to the case of formation damage, the clay particles were not released because the salinity level was sufficiently high to keep the clay particles in their original positions (Lager et al. 2006, 2007). This suggests that if the chemistry of injection water is manipulated sufficiently closely, it could be possible to achieve a wettability modification without inducing clay-related impairment problems.

Lager et al. (2006, 2007) found that after ageing core material in the presence of highly saline NaCl brine and subsequently flooding with high-salinity NaCl brine, incremental oil production was obtained when compared to a case where the ageing and subsequent flooding was conducted using a brine that contained multivalent cations. This suggested that during the ageing with pure NaCl brine, a more water-wet state had been achieved because fewer hydrocarbons were adsorbed to clay surfaces, and it can be inferred that using high-salinity connate brine containing calcium and magnesium ions during ageing promotes oil-wetness through the adsorption of petroleum heavy ends on clay mineral surfaces. Consequently, this would result in a relatively poor oil recovery.

The results suggest that the injection of low-salinity water changes the chemical environment of the adsorbed petroleum heavy ends on clay mineral surfaces, from one with a high concentration of divalent cations to one with a low concentration of divalent cations. Because of the cation-exchange processes, fewer divalent cations are available to keep oil attached to the clay surfaces.

However, the cation-exchange processes on their own are not sufficient to improve oil recovery. While the cation exchange is occurring, the charge density at the surface of the clay minerals decreases. By lowering the salinity, the electric double layer of the clay minerals expands, and increased electrostatic repulsive forces allow mobilization and, hence, the actual removal of the negatively charged oil components from the negatively charged clay mineral surfaces. (The charged surface of a rock or clay attracts ions of the opposite charge in the water phase, and the thickness of this double layer is a function of the concentration of ions in the water phase and the charge of the ions.) Basically, the injection of low-salinity water leads to the removal of polar organic components from the clay surfaces, replacing them with uncomplexed cations. This eventually results in increased water-wetness and additional oil recovery.

Lager et al. (2006, 2007) also suggested

- The major factors influencing the low-salinity effect are the clay content of the rock, the composition of the connate water, and the composition of the injected brine. A linear correlation between clay content and oil recovery was suggested.
- Different clays could have different impacts—swelling clays are detrimental because of the potential for formation damage induced by fresh water. Furthermore, the application of low-salinity flooding could have no effect in the presence of positive zeta-potential (i.e., positively charged) clay minerals such as chlorite.
- To trigger the cation-exchange process, the injection brine should have not only low salinity but also a multivalent-cation concentration that is lower than that of the connate brine. In other words, a low-salinity brine with a high calcium content would be much less likely to generate a low-salinity recovery benefit.
- A brine salinity of approximately 3,000 ppm TDS seems to be appropriate for a strong low-salinity effect, whereas a 6,000-ppm-TDS brine represents the upper threshold for a positive effect.

It appears, therefore, that the extent of the improved oil recovery from wettability modification depends not only on the reduction of the level of cations present but also on the amount and nature of the clay minerals coating the rock, and especially on their distribution over the rock surface. It has been specifically suggested that kaolinite could be wetted by crude oil and therefore be of benefit in generating increased oil recovery from low-salinity flooding (Jerauld et al. 2008).

Despite the strong evidence presented regarding the low-salinity mechanism in all the work described so far, a number of other potential mechanisms have been proposed:

1. Reduction in interfacial tension between water and oil phase: McGuire et al. (2005) observed reductions in interfacial tension between the water and oil phases of approximately 30%. However, a reduction in interfacial tension of several orders of magnitude would be needed to have a significant

impact on the relative permeability curves (and, hence, on oil production), which negates the likelihood that this could be an important mechanism. In laboratory-scale core flow experiments, where the capillary pressure dominates the production behavior, the occurrence of incremental oil production by means of some lowering of the water/oil interfacial tension cannot be excluded. Hence, this issue needs careful attention during the evaluation of laboratory data.

2. Reduction of remaining oil saturation by increased pressure differential: A coreflood study identified that in the presence of water-sensitive clays, it was possible to produce more oil than brine as a result of the development of a high differential pressure (Bernard 1967). Only a limited differential pressure can be applied over a small laboratory core, and consequently, a typical mixed-wet core can contain only a limited saturation range. By increasing the differential pressure over the core, a larger range of water saturations can be contained in the small core plug. This implies that, although a differential-pressure increase over a mixed-wet core will lead to additional oil production, this is caused by the important role of the capillary pressure on the core-plug scale.

   On the reservoir scale, the capillary pressure gradient term can normally be ignored in the fractional flow because of the small saturation gradients. Therefore, increasing the differential pressure over the formation on a reservoir scale is not expected to lead to any significant additional oil production. This mechanism can therefore be discounted as an explanation for the low-salinity-flood effect because sensitivities of relative permeability curves and residual oil saturations to rate and pressure differential that have been found in laboratory tests almost certainly result from capillary end effects and the saturation anomaly near the outflow boundary of a small core plug.

3. Water-diversion mechanism: Cation-exchange processes cause modified ionic compositions that can cause swelling (montmorillonites/smectites) or deflocculation (kaolinites, illites). After the release of fines, these compositions will be carried farther into the reservoir by the injected water and can subsequently block pore throats. When established flow channels become completely or partially closed, new flow channels might be established and flooded. A combination of pore-throat blocking, the plugging of flow channels, and an apparent water-phase viscosity increase from the suspended material delivers a reduced mobility of the water phase. The reduction in oil-phase mobility is much less than that of the water phase, and as a result, the oil rate will be accelerated with respect to the water and will be produced at a lower water cut compared to normal waterflooding.

   Bernard (1967) suggested that montmorillonite swelling could have a similar effect, with the reduction in pore space possibly leading to additional oil production. The swelling leads to a reduction in water-phase permeability and to a lower mobility of the water phase compared to normal water. For reasons similar to those described previously, the oil phase is again accelerated with respect to the water and will be produced at a lower water cut than in normal waterflooding.

   Although this diversion mechanism might sound plausible, it must be recognized that in diverting water from the higher-permeability intervals in this way, the water is then diverted to lower-permeability intervals in which there is presumably a similar amount of clay present. Consequently, the plugging mechanism in the sand to which the water has been diverted presumably occurs much more quickly, and therefore, the benefits from such a mechanism might be small.

Based on experiments conducted by various groups investigating low-salinity-flooding effects in sandstones, the following conclusions can be reasonably drawn:

- The mechanism for increased oil recovery from low-salinity flooding in sandstones is most likely a wettability modification toward a more water-wet state, with increased oil recovery occurring through the improvement of the microscopic sweep efficiency.
- In a water-wet reservoir, low-salinity flooding is not expected to yield additional oil production by means of wettability modifications. It could be an appropriate technology to consider for reservoirs where the wettability is mixed-wet to oil-wet and where there is an appropriate clay mineralogy present in the reservoir. Montmorillonite clays are considered unfavorable for low-salinity flooding because of the dangers of clay swelling. The presence of kaolinite is considered particularly favorable for low-salinity flooding, but low-salinity flooding could also possibly be applied with illites and mixed-layer clays. It is expected to be ineffective in reservoirs that contain an abundance of freshwater-insensitive clay minerals such as chlorites.
- The process is probably most efficient when applied from the outset of a waterflood. It could still generate incremental oil when applied in tertiary mode, but the incremental oil will come as a small slither that could be less economic than in cases where low-salinity flooding is applied from the outset.
- Low-salinity flooding targets wettability modification to create a more water-wet state, expelling adsorbed hydrocarbons from fine clay particles by means of increased electrostatic repulsive forces. The salinity level must be closely controlled and should be sufficiently low to allow for increased electrostatic repulsive forces by double-layer expansion and sufficiently high to avoid formation damage.
- The ideal injection-water chemistry will be critically dependent on the clay mineralogy present, but the available data suggest that appropriate salinities range from 3,000 to 6,000 ppm TDS. The removal of multivalent cations from the injected water might also be important because these could be activators for hydrocarbon adsorption to the clays and could effectively suppress double-layer expansion.
- On the basis of the observed mechanism, it is possible that in some conventional floods, a "reverse low-salinity effect" occurs. For example, if seawater is injected into reservoirs with very fresh formation waters, it can be speculated that the high level of multivalent cations in seawater induces a change to a more oil-wet condition, thereby jeopardizing sweep and recovery.
- Theoretically at least, an additional improvement to a low-salinity flood might be achieved by the suppression of flood-front instabilities by the addition of a polymer to the low-salinity water.

Because of the current immature state of field deployments, there is appreciable uncertainty regarding the extent to which benefits reported from small-scale core-based studies will translate to field-scale recovery using this technology. Nevertheless, assuming that some benefits are present, there could be other material supplementary benefits from low-salinity-flooding deployment. These might include

- A reduced reservoir-souring risk resulting from the removal of sulfate ions
- Reduced scaling risks, also associated with the removal of sulfate ions
- Improved injectivity because the facilities needed for ion removal demand low solids content to avoid plugging in the membrane filtration process
- Delivery of a gateway to chemical enhanced oil recovery (EOR) processes, reducing chemical/polymer requirements for those processes.

5.2.1 Timing for Low-Salinity-Flood Applications. A low-salinity flood in secondary mode corresponds to the case where low-salinity water is applied from the outset of injection, while a low-salinity flood is said to have taken place in tertiary mode when it occurs after a period of injection using a conventional water source that confers no recovery benefit. These terms are based on the way that waterflood was historically applied after an initial depletion phase. In such a scenario, water injection became a secondary recovery process, and subsequent EOR schemes were then tertiary recovery processes. In reality, a low-salinity flood applied from the outset is a primary process, and when it is applied after a conventional waterflood, it is usually a secondary recovery process. However, because of the way that waterflood was historically deployed, they are commonly referred to in the literature as secondary and tertiary low-salinity floods, respectively. Therefore, to avoid confusion, this section will use the terminologies typically used in the literature, referring to a low-salinity flood applied from the beginning as a secondary process and one that is applied after a conventional waterflood as a tertiary scheme.

In a secondary waterflood, the change in wettability is described by an extended Buckley-Leverett theory (Jerauld et al. 2008). **Fig. 31** shows the anticipated water-saturation profile for an initially oil-wet system as a function of the distance from the injection well for high-salinity injection (blue) and low-salinity injection (purple). When measured in terms of cumulative injection, low-salinity flooding will delay the moment of water breakthrough and will also slow down the increase in water cut. Consequently, it can be observed that low-salinity flooding provides the same cumulative oil recovery at a lower water cut or a higher cumulative oil recovery at the same water cut. It can also be observed that there is oil banking ahead of the low-salinity shock front as a result of the accumulation of de-adsorbed oil, leading to constant water saturation over a finite distance, and that there is a reduced oil saturation remaining behind the shock front.

Beginning a low-salinity flood after a period of conventional waterflooding (tertiary low-salinity flood) results in the mobilization and production of a tertiary oil bank because oil that was residual oil with respect to conventional water injection becomes mobilized after contact with the low-salinity water. At the producing wells, this should be characterized by a temporary drop in water cut, immediately before breakthrough of the low-salinity water.

Any delay in switching to low-salinity water will lead to a smaller volume of oil that could be accelerated. This reduces the duration of production of the tertiary oil bank and thus reduces the beneficial effect on cumulative oil production.

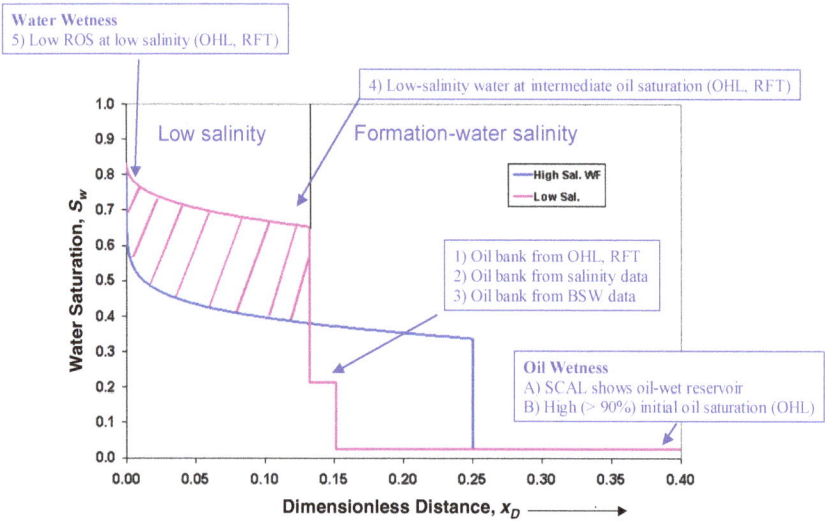

Fig. 31—Saturation profile for a conventional waterflood (WF) and a low-salinity water-flood (Vledder et al. 2010). SCAL = special core analysis, ROS = residual oil saturation, OHL = openhole log, BSW = basic sediment and water, and RFT = repeat formation test.

Consequently, the application of low-salinity flooding at an early stage will maximize the amount of movable oil for which the production can be accelerated. The cumulative-oil-production profile with low-salinity flooding from the moment of breakthrough of the oil bank onward is identical for both modes. However, in tertiary mode, the oil-bank breakthrough takes place later in time and, hence, later by amount of injected water compared to secondary injection mode. During this injection period, oil production by conventional waterflooding has continued. Therefore, the incremental oil recovery with respect to conventional waterflooding is reduced in tertiary mode. This view is supported by experiments (Fjelde et at. 2013) that suggested a faster recovery for low-salinity flooding in secondary mode.

5.2.2 Potential Application Environments. It should be stressed that the extent of improved oil recovery from wettability modification depends not only on the reduction of the concentration of cations present but also on the amount and nature of the clay minerals coating the rock, and particularly their distribution over the rock surface. Furthermore, the types of clays that are present in the reservoir will also be important. It has previously been noted that low-salinity injection can induce clay swelling and/or fines-mobilization problems. Because this should be avoided, the implication is that the salinity of the low-salinity water needs to be low enough to generate a positive recovery benefit but not so low as to induce clay problems. Because montmorillonite is the clay that is the most sensitive to a salinity shock, reservoirs that contain this clay might not be the ideal candidates for application. On the other hand, kaolinite is less sensitive to such problems, and reservoirs containing kaolinite might represent ideal candidates for low-salinity applications.

Other relevant factors include water/oil gravity drainage, initial water saturation, clay distribution, clay type, salinity level of the formation water, availability of a natural low-salinity water source, and oil composition.

Water/Oil Gravity Drainage. In reservoirs displaying significant water/oil gravity drainage, the amount of target oil for low-salinity flooding might be relatively small.

Initial Water Saturation. In the transition zone, fewer of the pores will be oil-filled. As a result, although there could still be a response to low-salinity injection, the magnitude of the response will be lower than it would have been if the initial oil saturation were higher, making the technology less attractive for such situations.

Clay Distribution. Clay minerals can be present as finely dispersed particles covering the rock surface and as aggregates within the pore space. If the clays are covering the rock surface, they will make it more oil-wet, making low-salinity flooding more attractive. This suggests that scanning electron microscopy analysis on thin sections of the core plugs could be useful to assess both the types of clays and their distributions.

Clay Type. The mechanisms suggest that reservoirs in which the clay content is dominated by kaolinite are appropriate low-salinity-flood candidates. Conversely, the presence of montmorillonite is likely unfavorable for low-salinity flooding.

Salinity Level of the Formation Water. More-saline formation brines usually equate to less water-wet conditions within the reservoir and thus could imply good scope for low-salinity flooding. However, mixing the low-salinity injection water and the connate water can increase the salinity of the water as it passes through the reservoir. This effect would potentially significantly increase the number of pore volumes of low-salinity water needed to obtain the required benefit throughout the reservoir. This effect should be factored in to determine the scope for this technology because in some cases, it can look attractive on a pore scale but end up being less attractive on a reservoir scale.

Availability of a Natural Low-Salinity Water Source. Some low-salinity-flood applications could require some degree of water treatment to generate the correct water chemistry to deliver an incremental recovery effect. However, in some cases, a natural low-salinity water source could already be readily available. In such cases, the incremental recovery is essentially free, making this a very attractive proposition. However, care needs to be taken because some naturally occurring low-salinity waters could have salinity levels that are too low, inducing clay-related problems. This might occur with river-water sources. On the other hand, if no montmorillonite is present within the reservoir rock, it is possible that there could be no negative factors associated with the use of very fresh river waters.

Oil Composition. The oil composition affects the oil's tendency to stick to reservoir rocks. This suggests that polar groups should be present in the oil to promote its ability to stick to the rock. The more such components are present, the more the oil is likely to wet the formation. The total acid number of the crude is an appropriate indicator of this parameter. Potentially, the residual oil saturation could also be a guide. In cases where the residual oil saturation is high, this could be the result of oil-wet characteristics of the rock and might suggest a large target for incremental recovery through low-salinity flooding.

5.2.3 Optimal Water Chemistry. To avoid clay-related problems in low-salinity floods, the salinity must be high enough to stay out of the danger area in the Scheuerman diagram (Fig. 29) for any clay types that are present in the reservoir.

This will be easier to achieve for cases where montmorillonite is absent because this clay type is the most easily damaged by salinity changes. On the other hand, anecdotal evidence suggests that the TDS content of the water needs to be reduced to less than 6000 mg/L to produce a reasonably large low-salinity effect.

Note, however, that low-salinity flooding also requires careful management of the divalent-ion content, as noted previously. Because of this requirement, a process that uses both nanofiltration and reverse-osmosis stages has been suggested (Ayirala et al. 2010). The nanofiltration process reduces the hardness of the water, including the removal of sulfate, with the reverse osmosis reducing the salinity. The output from such a process in series is water with very low TDS, raising the possibility of clay-related problems. Consequently, small quantities of the reject streams at each stage of treatment are blended back in to provide the required water quality specifications. Additional details regarding the facilities required to deliver the specified water quality for low salinity flooding are provided in *Waterflooding: Facilities and Operations*, another book in this series.

5.2.4 Field Experience. Despite a wealth of laboratory data, there is not always consistency regarding the extent of the effects as seen in those data. Part of the reason for that could be that core plugs are extremely small in comparison to the reservoir, and, second, the laboratory experiments do not perfectly replicate the field processes. Thus, although much of the laboratory test data do appear encouraging, field data are still needed to demonstrate that this technology merits further investigation and application.

It is somewhat surprising that, despite the enthusiasm shown by the industry in sponsoring laboratory-based studies, the field experience is very limited. A few "log-inject-log" experiments have been conducted in wells whereby a well is logged, a low-salinity slug is then injected, and the well is subsequently relogged to ascertain the effects of the low-salinity slug. Webb et al. (2004) describe a log-inject-log test in a producing well, where 10 to 15 pore volumes of seawater was first injected to derive a baseline residual oil saturation, followed by sequences of more dilute brine injection, with the results suggesting a reduction in residual oil saturation between 25 and 50%. Single-well chemical tracer tests have also been conducted in three fields in Alaska, and these also suggested reductions in residual oil saturation that were reported as translatable to incremental recoveries between 6 and 12% of the original oil in place (OOIP) (McGuire et al. 2005).

The problem with such tests is that the number of pore volumes injected at the wellbore is much larger than the number of pore volumes injected in the reservoir, and, as such, the results of these tests are unlikely to represent the size of the effect in the field. These tests are therefore able to demonstrate that the low-salinity effect is real, but they still do not show that low-salinity flooding is a viable economic technology on a fieldwide scale.

Following single-well tests in the Endicott Field in Alaska (Seccombe et al. 2008), work was conducted to evaluate if a low-salinity effect could also be observed at the interwell scale (Seccombe et al. 2010).This trial was designed to evaluate if mixing reduces the improved recovery effect and if the adverse mobility ratio between the injected water and the oil bank causes viscous fingering. This test was implemented in a single reservoir zone using an injector and a producer that were 317 m apart. The clay content in this zone was 12%, with kaolinite the dominant clay followed by illite, and the zone had a residual oil saturation to high-salinity injection of 41%, suggesting a good low-salinity-flooding candidate. The producer was monitored for changes in water cut and ionic composition.

At first, produced saline water was injected to flood the pattern to a water cut of more than 95%. Reduced-salinity water injection then began, and after approximately 2 months of low-salinity water injection, an increase in oil rate and a reduction in water cut were observed, indicating the mobilization of an additional oil bank, with the increase in oil rate immediately followed by the arrival of reduced-salinity water (**Fig. 32**).

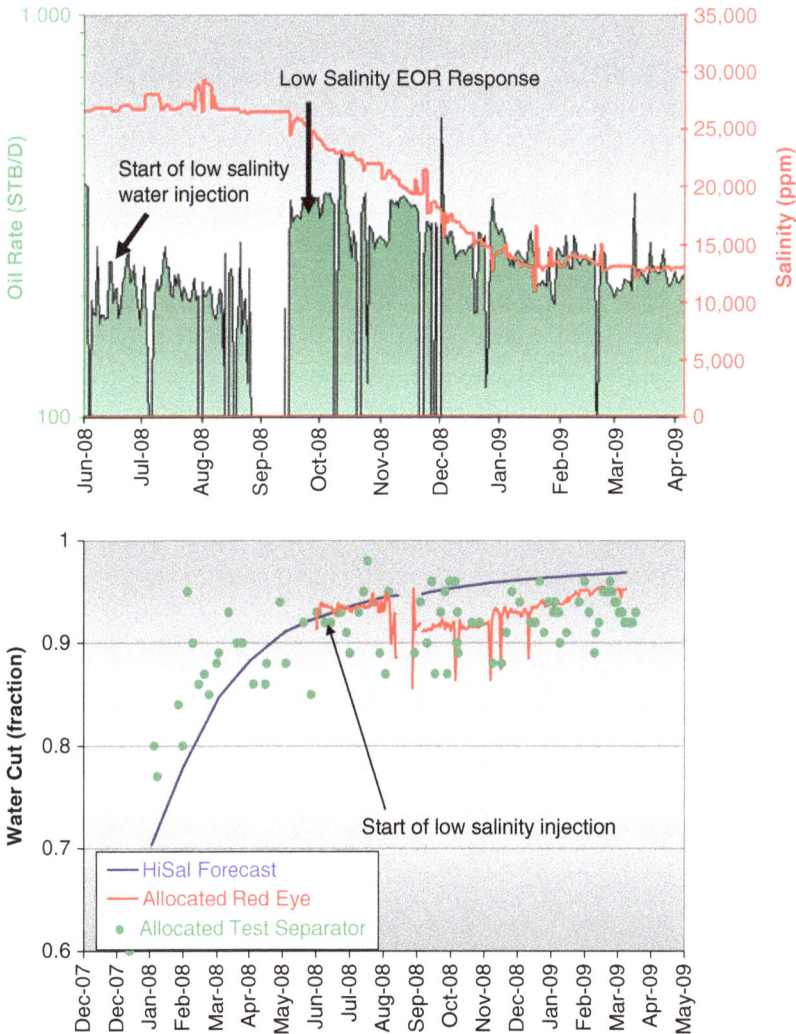

Fig. 32—Oil-rate and water-cut response in Endicott low-salinity pilot (Seccombe et al. 2010).

The oil response was in line with that predicted by corefloods and single-well tests, but the associated reduction in producing water cut, from 95 to 93%, was less than had been expected. Analysis of the produced-water composition showed that 45% of the water was coming from outside the pilot area. Backing out the associated production from outside the pilot area, it was estimated that the effective drop in water cut within the pilot area was 5.5%.

Analysis of the results of this test compared to predictions of performance if high-salinity injection had been continued suggests an incremental recovery of 10% of the OOIP. However, as can be observed from the accuracy of the water measurements and their scatter in Fig. 32, there could well be a significant degree of uncertainty associated with this assessment.

Another fieldwide response has reportedly been observed in the Omar Field in Syria (Vledder et al. 2010). Water injection began in this field in 1991 using fresh water taken from the Euphrates River (salinity of 500 mg/L and very low bivalent-ion content). The formation water has a salinity of 90 000 mg/L, with a high bivalent-ion content of 5000 mg/L. The clay content is 0.5 to 4%, nearly all of which (95 to 100%) is kaolinite. After the injection of approximately 0.4 pore volumes of this fresh water by 2004, injection was converted to produced water. Wettability changes have been observed in the field, and a stepwise change in the water cut of some producing wells was seen (**Fig. 33**), which was said to be indicative of the change in wettability predicted by the Buckley-Leverett theory. Vledder et al. (2010) report that the change could be responsible for an incremental recovery of approximately 10 to 15% of the OOIP. It should be noted, however, that not all producers in this field demonstrated the water-cut development attributed to the low-salinity effect. Furthermore, such a characteristic profile might also be observed where water breakthrough occurs in two separate intervals. As such, the claims that this case represents a field-scale demonstration of incremental recovery associated with low-salinity water injection might not be as compelling as suggested.

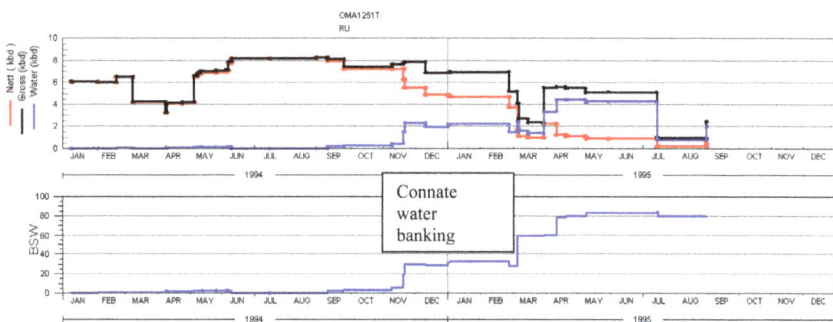

Fig. 33—Oil and water production (top) and water-cut development (bottom) for OMA125 (Vledder et al. 2010). BSW = basic sediment and water.

Many fields in the Powder River Basin of Wyoming have been flooded with water from low-salinity sources (Robertson 2007). However, many of the available production data are weak. The waterflood responses in three Minnelusa Formation fields (West Semlek, North Semlek, and Moran), where good data were available, were analyzed, with ultimate recoveries from the fields plotted as a function of the ratio of the average salinity of the injection water to the salinity of the formation water (**Fig. 34**). This was reported to indicate a trend demonstrating a higher recovery factor with a lower salinity ratio. However, the evidence again looks rather weak because polymer injection took place in all three fields. Furthermore, it is not known if the differences in recovery might be explained by other factors, such as differences in sweep efficiency in the three fields. If all other factors were known to be the same in each field, the results would confirm and support the positive low-salinity effect observed in laboratory data.

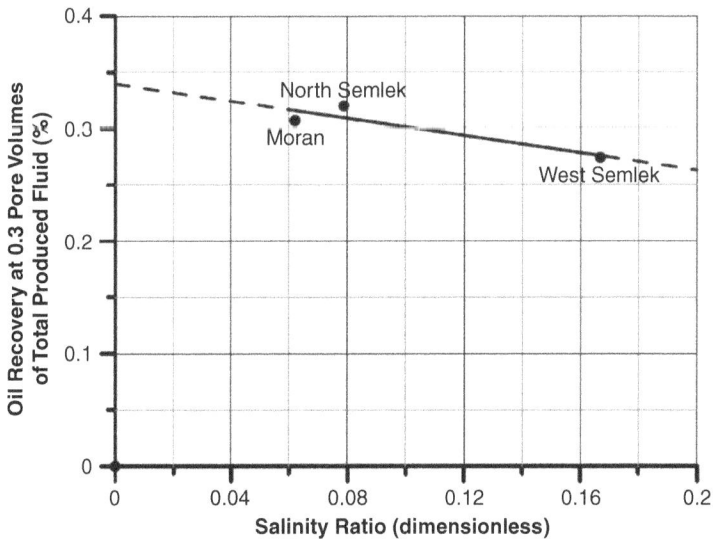

Fig. 34—Relationship between oil recovery and salinity ratio for three Minnelusa waterfloods (Robertson 2007).

To date, no project has injected water in which the composition was specifically tailored to create a low-salinity effect from the outset, although this is the plan for Phase 2 of the Clair development west of Shetland in the UK (Robbana et al. 2012). Overall, it is estimated that low-salinity waterflooding in Clair Ridge will produce 7% more OOIP than conventional seawater flooding at a development cost of USD 3 per barrel. Conceptually, that should be the best time to perform a low-salinity flood compared with a later (tertiary) application, when more water would be expected to be needed to deliver a smaller incremental slither. However, a drawback of the early deployment is that it becomes a leap of faith, with the benefits more difficult to quantify.

Although the low-salinity effect does appear to be real, the results from laboratory experiments are varied, and it is difficult to relate the results from those laboratory conditions to field performance. For example, although most agree that wettability plays a significant role in low-salinity-flood effects, there is no full consensus on the exact wettability change needed to generate beneficial effects. This could be because core-based experiments are not always adequately robust in the design of the conditions under which they are run. Wettability is a dynamic, not a fixed, parameter, and it can be changed when key parameters are changed (Skrettingland et al. 2010). Increases in recovery can be observed as the wetting conditions change, and these might or might not relate to the conditions that would be observed in the reservoir during a real waterflood.

This perhaps explains why there has been little adoption of the technology despite an enormous amount of interest, as evidenced by the number of studies. Therefore, the initial field deployments are most likely to occur in cases where there could be secondary benefits such that the incremental recovery benefits are not central to the justification. The reduction or removal of materially significant scaling and souring problems might potentially provide those drivers for some specific applications.

5.2.5 Low-Salinity Flooding in Carbonate Reservoirs. Data suggest that it might also be possible to manipulate injection-water chemistry to optimize recovery in carbonates, although much less work has been conducted to assess the impact of low-salinity flooding in carbonate rocks compared to the number of studies of clastics. Consequently, the details of the process are less understood for carbonates than they are for sandstones. Early studies on wettability modification in carbonate reservoirs were conducted at the University of Stavanger, in Norway. Austad et al. (2005), in research conducted on chalk outcrop samples, suggested that seawater, or modified seawater, provides the ideal chemistry to optimize oil recovery compared to other types of water.

Austad et al. (2005) found that although wettability changes appear to again be involved, the mechanism is fundamentally different from that seen in sandstones. Carbonates are commonly observed to be more oil-wet than sandstones because at the pH levels present in reservoirs, the rock carries a positive charge, thereby making it attractive to carboxylic-acid groups (COO$^-$) present in the crude oil. Hence, it is logical that a reduction in the oil-wetting character will improve the flood efficiency. Because the quantity of carboxylic acids in the crude oil is strongly indicated by the total acid number of the crude, this suggests it is a very important wetting parameter for carbonates.

Although there are strong indications of the role of wettability modification in the process, as is the case with sandstones, a number of different mechanisms have been suggested. These include fines migration, rock dissolution, a reduction of interfacial tension, microemulsion formation, an expansion of the electric double layer, and ion exchange. Austad et al. (2005) suggest that the quantities of calcium, magnesium, and sulfate ions are critically important (**Fig. 35**). As the rock is flooded by seawater, sulfate ions in the seawater can bind to the carbonate surface and displace polar oil components from it, thereby increasing its water-wetness. The role of calcium ions is that they carry a positive charge, and so they can attract more negatively charged sulfate ions when the latter are present at the surface. If more sulfate is present near the surface, more calcium will also be drawn close to the surface to be used in substitution reactions. Consequently, sulfate can have a dual role—it can take the place of a carboxylic-acid group on the calcium rock site, and it also reduces repulsion, enabling the calcium to form complexes with the carboxylic acid.

At high temperatures (greater than 100°C), magnesium ions can also participate in the reaction. Because the magnesium ion is strongly hydrated in water at low

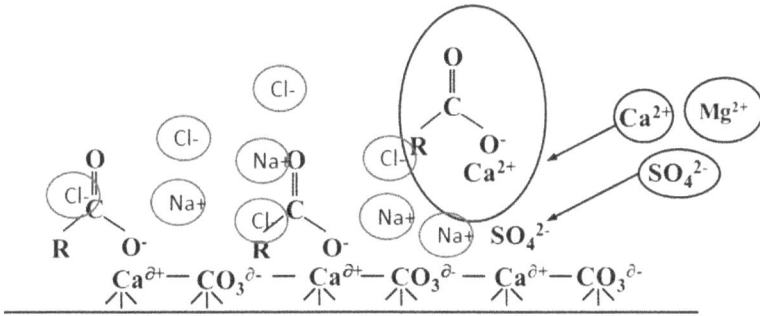

Fig. 35—Low-salinity mechanism in carbonate reservoirs.

temperatures, its ability to react with other minerals is impaired. As the temperature increases, the quantity of water molecules in the magnesium hydration shell is reduced, and the ion reactivity is thus increased.

Jafar Fathi et al. (2010) reported that oil recovery could be improved still further by removing or reducing the concentrations of sodium and chloride ions in the seawater. Chloride ions present in the diffuse outer layer repel negatively charged sulfate ions, so lowering the NaCl concentration in the solution could further promote sulfate adsorption onto the surface. Diluting the seawater and, hence, reducing the concentration of all ions had an adverse effect on the benefits.

Studies confirm the ability of sulfate to act as a wettability modifier toward the chalk surface in experiments in which live crude oil and brine were used. Webb et al. (2005) investigated the impact sulfate ions had on the imbibition capillary pressure and saturation changes using two identical chalk cores from the Valhall Field, in the Norwegian North Sea. The results support the view that the wettability of the Valhall core material changed toward a more water-wet state in the presence of sulfate.

By contrast, however, Ferno et al. (2011) measured additional oil recovery in spontaneous imbibition tests from aged outcrop core plugs placed into brine with and without sulfate. Core plugs from three different quarries (Stevns, Rordal, and Niobrara) were prepared under similar conditions, and it was found that the effect of sulfate was dependent on the chalk type because increased oil recovery was observed only in the Stevns outcrop chalk. The difference in behavior was attributed to compositional differences between the different outcrops.

Strand et al. (2008) reported that similar recovery benefits can also be observed in limestones. The imbibition behavior of two identically prepared limestone plugs was compared—one plug was imbibed with seawater, and the other with seawater without sulfate. The oil recovery from the plug immersed in the seawater was approximately 15% higher than that from the plug immersed in the seawater without sulfate.

Anhydrite mineral dissolution has been proposed as a possible explanation for the incremental recovery observed after the injection of low-salinity brine in carbonate (Pu et al. 2010) on the basis of coreflood experiments performed on core from two clastic formations and one dolomitic formation. In the dolomitic cores, the injection of a diluted reservoir brine with 1,538 ppm TDS resulted in an incremental oil recovery in the range of 6 to 8% of oil-in-place. Anhydrite dissolution was evident from the increase in the sulfate-ion concentration in the core effluent.

Perhaps the most compelling work on carbonate systems has been conducted in Saudi Arabia. Coreflooding studies were performed under reservoir temperature conditions using live crude and in composite cores so that production resulting from the reduction of capillary end effects would be small compared with the total recovery (Yousef et al. 2011). An additional oil recovery of 7 to 8.5% was obtained when twice-diluted seawater (TDS 28,835 ppm) was injected; an incremental 9 to 10% was observed with 10-times-diluted seawater (TDS 5,767 ppm); and an incremental 1 to 1.6% was observed with 20-times-diluted seawater. Interfacial-tension measurements were performed during the study, and it was observed that as the salinity of the injection brine decreased, the interfacial tension decreased. However, a significant reduction in interfacial tension was observed only when field connate water (TDS 213,734 ppm) was replaced by regular seawater. Negligible impacts on interfacial tension were observed with the different dilutions. Trends in contact angle measurements supported the view that the cause of the additional oil recovery was wettability alteration.

The same group also conducted the first field-scale test in carbonates by performing two single-well chemical tracer tests in a Jurassic carbonate reservoir using a diluted seawater (Yousef et al. 2012a). The tests indicated an approximate 7-saturation-unit reduction in the residual oil beyond conventional seawater injection. Yousef et al. (2012b) performed an additional study designed to better elucidate the mechanism(s) at play through the sequential injection of various seawater dilutions following an initial phase designed to flood the core to residual oil saturation (**Fig. 36**). Rock-surface chemistry studies from the coreflood work using

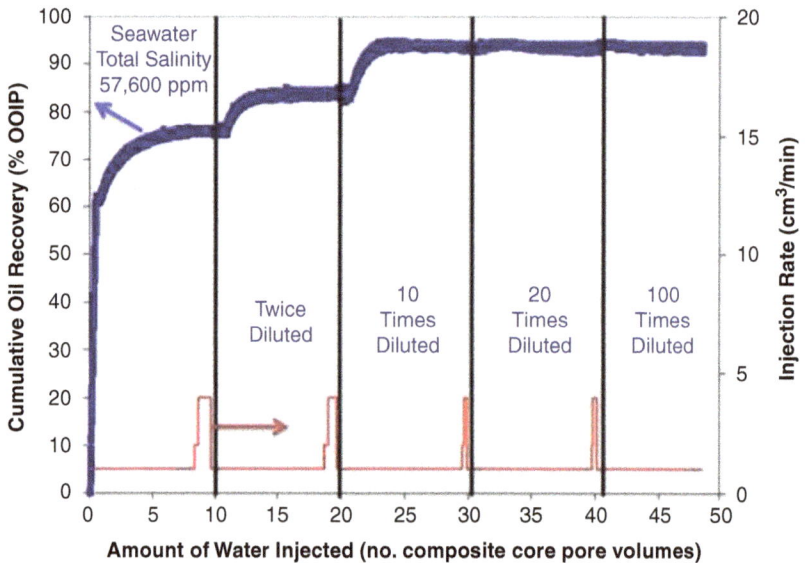

Fig. 36—Oil-recovery curves from reservoir coreflood experiments with seawater dilutions (after Yousef et al. 2012b).

nuclear magnetic resonance spectroscopy confirmed that injecting different salinity slugs caused changes to the surface charges on the rock, leading to more interaction with water molecules, suggesting that changes to the surface charge are at least one of the mechanisms at play. Zeta-potential results showed that diluted seawater has the ability to make the surface of the rock more negatively charged as a result of the dilute-brine injection. Changes in contact angle showed that multivalent ions might play an important role in the wettability change and that the process is therefore not a straightforward low-salinity process.

The impact of ionic brine composition on oil recovery has been evaluated through numerous corefloods using dolomite-outcrop and limestone reservoir-rock materials (Gupta et al. 2011). The impact of softened formation water, seawater (with and without sulfate), and seawater with added phosphate and borate ions was evaluated. Gupta et al. (2011) demonstrated that adding borate or phosphate instead of sulfate to the seawater resulted in a higher incremental recovery as well as a more rapid oil response. The wettability modification in this case might have occurred as a result of phosphate adsorption.

Several low-salinity waterflood experiments have been conducted using different carbonate cores (Al Harrasi et al. 2012). Both coreflooding and spontaneous-imbibition experiments were conducted using synthetic brine at 2-, 5-, 10-, and 100-times dilutions, yielding increases in oil recovery of 16 to 21%.

In work conducted using live oil (Zahid et al. 2012), no incremental recovery benefits were found at ambient temperatures, although an increase in pressure drop was observed. However, an increase in oil recovery was observed with runs at high temperatures (90°C), perhaps suggesting that magnesium could be an important contributor to the benefits observed in these experiments. This is because at elevated temperatures there is the potential for calcium ions on the rock surface to be substituted by magnesium ions.

In summary, the literature suggests that the mechanism of wettability alteration in carbonates is very different from that observed in sandstones and that it might be achieved through the combined action of calcium, magnesium, and sulfate ions driven by changes in surface charge and/or ion-exchange processes. Monovalent ions such as sodium and chloride have not been found to aid in this process because they might hinder the reactions of the active ions. A reduction in the sodium and chloride ions in the water will expand the electric double layer, increasing contact between the active ions (sulfate, calcium, magnesium) and the surface rock. However, the literature shows that there is much more uncertainty regarding carbonate wettability-modification mechanisms—the mechanism could be different depending on the rock type.

6. Anhydrite Swelling

Many carbonate reservoirs are associated with overlying salt pans or updip sabkhas that can contain gypsum (calcium sulfate dihydrate, $CaSO_4 \cdot 2H_2O$). As the rock undergoes the burial process over geological time, it is possible that the water molecules in the gypsum structure will be removed, thereby converting the gypsum into anhydrite:

$$CaSO_4 \cdot 2H_2O = CaSO_4 + 2H_2O. \quad\quad\quad\quad\quad\quad\quad\quad\quad\quad (15)$$

Alternatively, anhydrite can form through the evaporation of seawater. Massive accumulations of anhydrite occur when salt domes form a caprock. Anhydrite is typically 1 to 3% of the salt in salt domes and is generally left as a discrete layer when the halite is removed by pore waters, so the cap to a carbonate sequence is often anhydrite.

In addition to anhydrite layers above carbonate reservoirs, there can be a significant amount of anhydrite present within a carbonate sequence. This is often seen in dolomite reservoirs because the dolomitization of calcite by seawater results in an increased calcium concentration in the fluid and the precipitation of gypsum.

When water is injected into the reservoir as part of a waterflood, it is sometimes possible for the anhydrite to convert back to gypsum. This change carries with it a swelling risk because the reintroduction of water into the calcium sulfate structure brings with it a volumetric increase. The molar volumes of anhydrite and gypsum at standard conditions are 46.11 cm³/mol and 74.15 cm³/mol, respectively, and because only small density changes might be attributed to temperature and pressure variations, a volumetric increase of approximately 60% associated with the transition is implied.

When the anhydrite is present as the overburden, this is not a problem, but anhydrite dispersed within the producing reservoir needs to be assessed for its swelling potential because of impairment concerns.

Temperature is the most important control on anhydrite-to-gypsum transitions in a waterflood setting (Billo 1987). Solubility trends define the stability field of both minerals and transition temperatures, with anhydrite being the stable mineral in low-salinity waters at temperatures generally greater than 50°C. The anhydrite/gypsum transition temperature is also dependent on the NaCl concentration. The solubility of both anhydrite and gypsum increases with increased NaCl concentrations. In addition, increased NaCl concentrations favor anhydrite over gypsum stability, even at low temperatures. For cases in which swelling is a potential risk, control might be achievable through the appropriate selection of a water source that has relatively high salinity.

Anhydrite swelling has been encountered in the stringer reservoirs (a carbonate reservoir encased with a salt) of eastern Siberia. It is understood that a sour waterflood development in the Sredne-Botuobinsk Oil Field has resulted in very significant swelling in injection wells where freshwater injection was used, with the fresh water obtained either from the Cenomanian aquifer or from a river. Such swelling has required that many injectors be sidetracked.

Reactive transport modeling was conducted to assess the dangers of anhydrite swelling (as well as halite scaling) as part of waterflood studies on a carbonate stringer reservoir in Oman (Al-Mayahi et al. 2012). This showed that anhydrite/gypsum conversions were unlikely to result in injectivity or well-integrity problems and that, as a result, heating of the injection water would not be needed. This finding saved the project USD 160 million in operating costs.

7. Waxes

Wax deposition is commonly induced as a result of temperature changes, when the wax, which is a component of the crude oil, begins to come out of solution because the temperature has dropped below the wax-appearance temperature. Many fields could therefore suffer wax-deposition problems in the production process as a result of the temperature drops. As far as water injection is concerned, wax-deposition problems can occur because the injection water is invariably colder than the reservoir temperature.

When the wax-appearance temperature is equal to the reservoir temperature, so that any reduction in temperature can begin to induce problems, wax deposition can occur in the near-wellbore region of injectors where the injection of cold water causes the deposition. Such deposition can block pore throats near the injector, thereby resulting in a significant reduction in injectivity. Alternatively, in heterogeneous reservoirs, the wax might preferentially plug lower-permeability intervals and thereby have a negative impact on sweep even if the injectivity is not affected.

Such problems can be very difficult to overcome. The extent of the problems is indicated by the fact that the most commonly used treatment is to apply heat to the injection water—an expensive but sometimes necessary solution. This solution has been applied in Russia and to the Uzen Field in Kazakhstan. An application has also been reported in the Senex Field in Alberta, Canada (Cassinat et al. 2002).

8. Asphaltenes
Unlike that of waxes, asphaltene deposition is primarily driven by pressure changes.

Asphaltenes are the heaviest and most polar components in crude oil, but there is no universally accepted definition of what an asphaltene molecule is—the two most commonly used definitions are that it is the oil fraction that is insoluble in either pentane or heptane. There is no single asphaltene structure, but asphaltenes generally comprise a core of polyaromatic rings associated with a number of alkyl side chains.

Asphaltene behavior is somewhat complex, but these molecules might be partly dissolved in the crude oil and partly in a colloidal state in the oil, stabilized primarily by resin molecules that are adsorbed on the asphaltene surface. If this adsorption equilibrium is disturbed, for example, by the addition of a paraffinic solvent, the asphaltene particles will become destabilized and might then undergo an aggregation process to form larger particles.

In a waterflood setting, the injection of water helps to reduce pressure drops, and unlike that of waxes, asphaltene solubility is driven by pressure more than temperature. This is because as pressure drops, resin solubility decreases, and because resins help to stabilize asphaltenes, this can cause deposition in the near-wellbore region. Wang and Civan (2005) suggest that applying waterflood could help to avoid asphaltene-deposition problems that might otherwise be induced by depletion.

Counter to that, Yaseen and Mansoori (2018) suggest that injected water could induce asphaltenes to become less soluble in oil when water is partially misciblized in the oil phase. The driving force behind this asphaltene aggregation was shown to be an asphaltene/water hydrogen bond, and the effect was found to be more significant in cases where the injection water had very low salinity. However, the literature is effectively silent on the subject of injector impairment associated with asphaltene deposition, whereas there are a host of examples of production impairment resulting from asphaltene deposition as a result of pressure-drop effects. This suggests that, typically, a waterflood might be expected to mitigate asphaltene problems by means of stabilizing the reservoir pressure.

9. Hydrates

Hydrates are crystalline material formed between light hydrocarbon molecules and water in which the gas molecules become trapped in cages formed by hydrogen-bonded water molecules. Gas hydrates resemble ice, but, unlike ice, they can form at higher temperatures. This is because the presence of gas molecules provides extra attraction and, hence, stability, thus helping to fix the position of the water molecules and enabling freezing at higher temperatures.

Hydrate prevention is a key flow-assurance focus area, especially in deepwater fields. The conditions required to form hydrates are

- Free water (water in liquid form)
- Suitably sized small molecules such as methane, ethane, propane, *n*-butane, or CO_2
- Sufficiently high pressure, typically greater than 10 to 20 bar at ambient temperatures
- Sufficiently low temperatures, typically less than 20 to 25°C

Hydrate formation can block pipelines or wellbores, thereby stopping production or injection, so the conditions under which hydrates form need to be understood. Hydrate equilibrium curves can be calculated for a given composition using pressure/volume/temperature equation-of-state software, and hydrate prevention can be achieved by operating outside the thermodynamic hydrate formation envelope (temperature and pressure). Although there might be options to achieve this—such as by providing heat or insulation—there will be many cases in which this is not possible. In such cases, the injection of thermodynamic inhibitors such as methanol or glycol or the use of kinetic hydrate inhibitors will be required.

Fig. 37 shows a hydrate formation envelope and how it is increasingly constrained by increasing concentrations of methanol injection.

Because the presence of gas is a requirement for hydrate formation, most hydrate problems are encountered in the production system, and most attention inevitably focuses on this area. In water-injection wells, there definitely are no hydrate problems while the well is injecting water because all hydrocarbons are being pushed back into the reservoir away from the wellbore. However, when an injection well is shut in, there could be a danger of gas migrating back into the wellbore, and (high) pressure and (low) temperature conditions in water-injection systems are relatively conducive to hydrate formation if gas is present. This risk is much higher when wells are completed in the hydrocarbon column rather than in the water leg. Even then, the risks will be expected to diminish as more of the hydrocarbons are flushed away from the near-wellbore region, and as far as the injection wells are concerned, hydrate formation might be expected to be a relatively short-lived (potential) problem.

One option for control is to, if possible, set the subsurface safety valve below the point in the thermal gradient that forms hydrates, with a full column of water sitting on top of virgin reservoir pressure. The valve will be closed on every shutdown to limit the migration of gas above that point. The alternative is to provide for methanol injection at the tree and for this to be used generously when shutting down early in the injectors' life, before gas is largely pushed away from the well.

Eight cases of hydrate formation in water-injection wells have been reported (Rodrigues et al. 2009) in deepwater Brazilian fields, with each case requiring an

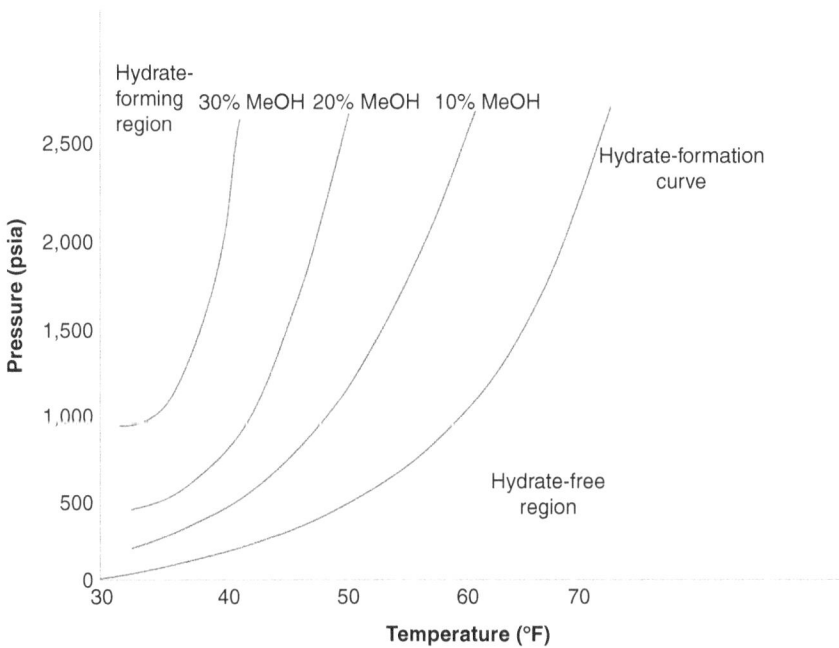

Fig. 37—Hydrate-formation envelope. MeOH = methanol.

expensive workover to remediate the well. Rodrigues et al. (2009) highlighted water hammer on shutdowns, well crossflow on shut-ins, and free reservoir gas segregation on shut-ins as potential means by which gas could re-enter the injection well, but they concluded that the latter effect was the one responsible for the problems observed in all eight wells. This analysis found that upon completion, wells were often left for months before starting injection, thus increasing the potential for hydrate formation. It notes, however, that not all wells suffered from hydrate formation—problems were most likely to occur in wells that were completed higher in the structure, close to a gas cap.

Rodrigues et al. (2009) doubted the value of a deep-set safety valve in controlling the problem because they are not designed to prevent gas segregation during a long shut-in period so they might not provide a seal against gas leakage, and, furthermore, the valve closure time would be too slow to prevent problems. Consequently, control is effected through the downhole injection of methanol or monoethylene glycol.

The removal of hydrate plugs is not easy (Freitas et al. 2002); hence, prevention is always preferable to cure.

10. Corrosion and Oxygen Control

It has previously been mentioned that severe localized corrosion problems can occur when H_2S is present. The overall level of corrosion of water-injection systems is usually driven by the presence of three dissolved gases—oxygen, CO_2, and H_2S. Of these gases, oxygen is usually the most corrosive; therefore, the control of corrosion often focuses on the control of this component, although CO_2 corrosion could be important in PWRI schemes.

10.1 Oxygen. Oxygen attack is proportional to the rate at which oxygen is transferred to the steel surface (as influenced by oxygen concentration and fluid velocity) and the surface temperature. It typically initiates as uniform corrosion on exposed steel surfaces. However, the corrosion products generated by that process might assist in passivating the surface such that subsequent levels of corrosion are reduced. This might seem beneficial, but sometimes localized pitting attack often initiates under corrosion scales as a result of the formation of differential aeration cells.

At dissolved oxygen concentrations less than 5 ppb, oxygen attack—uniform or pitting—is expected to be negligible.

It is often assumed that in PWRI systems, there will be no oxygen problems because oxygen is not present when the water is first taken from the reservoir and the water is confined in a pressurized system as it is treated at the surface. However, oxygen will diffuse in significant quantities into pressurized systems at any point where leaks are present (e.g., across pump seals, gaskets). Additionally, if any waterfloods are operated at low pressures such that an injection well is unable to support a full column of water when it is shut in, oxygen ingress will result.

Oxygen control is thus the primary methodology used for corrosion control in water-injection systems. It has previously been mentioned that negligible oxygen corrosion can be expected at a dissolved oxygen level of 5 ppb, but a practical limit for oxygen specifications is commonly set at 10 to 20 ppb. Some operators have operated with higher specifications of 20 to 30 ppb on the assumption that a passivating corrosion layer tends to offer protection at this level. Of course, it is always possible to provide for increased metal thickness to enable operation at slightly higher oxygen levels, but in waterfloods, corrosion is not the sole consideration.

It is important to recognize that corrosion processes generate products that add solids into the process that could have a very material impact on the solids loading. Because a great deal of effort is expended in removing solids in the main process plant, to avoid any negative impacts on injection-well injectivity and reservoir sweep, the general preference is to avoid any process that reintroduces such problems downstream of the plant. Most projects therefore tend to operate with tighter specifications of 10 ppb rather than rely on the creation of a passivating layer because there does not appear to be much logic in spending money to deliver a tight water quality specification only to reintroduce solids downstream of the plant. Much closer attention is paid to the oxygen specification in EOR projects because the combination of ferrous ions (Fe^{2+}) and oxygen is well-known to induce severe oxidative degradation to EOR polymers.

Fischer et al. (1996) reported on flow-loop tests to assess corrosion risks with oxygen concentrations between 20 and 200 ppb at temperatures between 6 and 13°C. The results suggested that for an oxygen level of 20 ppb, the corrosion rate of carbon steels will not be greater than 100 μm/yr, even at high flow rates. At oxygen levels between 20 and 50 ppb and at low flow velocities (2 to 4 m/s), the corrosion rates were less than 100 μm/yr, but at oxygen levels between 150 and 200 ppb, corrosion rates reached 200 to 1700 μm/yr.

Similar tests were conducted using a range of different metallurgies but at a temperature of 30°C (Schofield et al. 2004). Of the materials tested, only 22Cr duplex,

25Cr super duplex, and Alloy 718 offered "fit for life" solutions for service under the conditions of the test program (20-ppb oxygen with periodic excursions). Schofield et al. (2004) also identified that carbon steel and 1% Cr might be appropriate options provided that a finite life is acceptable.

The method of oxygen removal used is dependent on the amount of oxygen present in the stream. In seawater-injection systems, where the water is in equilibrium with the atmosphere, there are very appreciable amounts of oxygen present. Seawater treatment thus begins with a mechanical oxygen-removal process, often either a gas stripping or vacuum de-aerator tower. This typically delivers an oxygen specification of approximately 50 to 100 ppb. The reduction to the final oxygen specification is then achieved using chemical oxygen scavengers. For water sources where much smaller initial levels of oxygen are expected to be present (such as in the case of produced water), removal is normally achieved using chemical oxygen scavengers alone.

The most common types of scavengers are sulfite based [e.g., sodium bisulfite ($NaHSO_3$) and ammonium bisulfite (NH_4HSO_3)]. It is important to point out that although the scavengers are commonly supplied in the bisulfite form, the active scavenging species is the sulfite ion:

$$2SO_3^{2-} + O_2 \rightarrow 2SO_4^{2-}. \dots\dots\dots\dots\dots\dots\dots\dots\dots\dots\dots\dots\dots\dots\dots(16)$$

Because the ionic species distribution varies with pH, the scavenging effectivity is also pH dependent (Fig. 38).

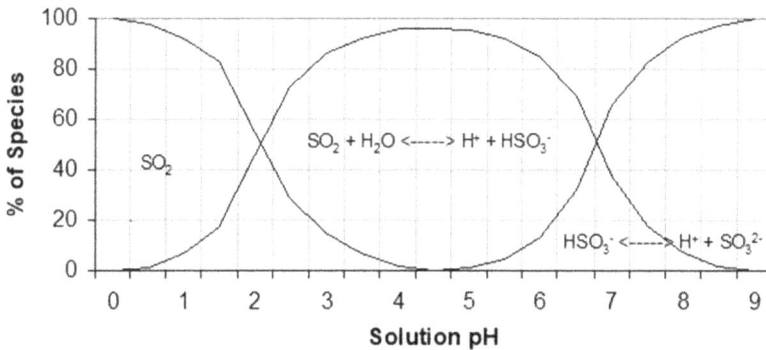

Fig. 38—pH dependency of sulfite.

From Fig. 38, it can be seen that at a pH in the range of 4.1 to 4.5 (which is the pH of commercially supplied 45 to 60% ammonium bisulfite or sodium bisulfite products), all the anionic species are present as bisulfite and therefore will not scavenge oxygen. This means that the product, as supplied, is stable with respect to atmospheric oxidation. However, if the product is stored in direct sunlight, it is possible that the elevated temperatures this can induce will promote decomposition, with the associated release of sulfur dioxide (SO_2) gas. This could either result in the release of the SO_2 or, if the vessel is sealed, induce an explosive release as the pressure in the vessel rises.

After the product is mixed with seawater in dilute solution (5 to 6 ppm will typically be added at the inlet to the retention section of a de-aerator), the pH of the mixture will normally be approximately 8.0 in vacuum de-aerator systems (and somewhat lower in gas-stripped systems, depending on the gas composition). At this pH, approximately 90% of the chemical is present in the reactive sulfite form and is therefore available for scavenging.

If a seawater-injection process relies entirely on a chemical scavenger for oxygen removal, this pH dependency is particularly significant because the very high dosage rates required (perhaps 100 to 200 mg/L) significantly reduce pH and will thus slow the rates of reaction. The total oxygen scavenging of cold waters is therefore particularly slow. This makes the chemical-only oxygen-removal process difficult to apply in lower-temperature environments such as the North Sea water. In fact, the problems are exacerbated because more dissolved oxygen is present in lower temperature seawater because the solubility of oxygen increases with decreasing temperature.

The reaction of bisulfite with oxygen is faster at higher temperatures, approximately doubling per each 10°C increase in temperature.

Transition metals have been found to be useful to kick off the reaction and to drive it to completion in a reasonable time frame. The product can therefore be supplied in either a catalyzed or a noncatalyzed form. (Small concentrations of transition metals present in seawater might conceivably already have performed the catalytic function.)

The two most common products used for oxygen scavenging are ammonium bisulfite and sodium bisulfite. Ammonium bisulfite has an advantage over sodium bisulfite in that it is soluble in a more concentrated solution (i.e., 65% w/w) at low ambient temperatures of approximately 5°C, conditions that would induce precipitation for the sodium salt. This makes the ammonium salt the preferred option for low-temperature applications (e.g., the North Sea) because it can be supplied in a more concentrated form, thus lowering volumetric requirements, making it preferable from an operational logistical perspective. In locations where ambient temperatures are higher, some operators prefer the sodium salt because there are concerns that the ammonium ion could be a potential nutrient for some bacteria. Additionally, the reaction rate has been reported as being slower for the ammonium salt compared to the sodium salt.

Where sulfite scavengers are used, it is common to run surveillance to detect a slight residual amount of scavenger, thereby ensuring that adequate scavenger has been injected. Concerns have been associated with this practice because excess scavenger could potentially lead to sulfide corrosion. In practice, however, the levels of excess scavenger that would be required to induce such problems are far in excess of those normally seen in injection systems, and this risk can be effectively discounted for most systems. Consequently, dosing to detect a slight excess of scavenger is a good practice, and it also ensures that there is no overdosing of the chemical; this can be particularly important because the reaction product is sulfate, which could increase the reservoir-souring risk in low-sulfate environments. Note, however, that because the reactions are not instantaneous, it is entirely possible that oxygen and sulfite will coexist; therefore, this should not be used in isolation to determine the required injection rate.

Another issue to consider is associated with the injection of hypochlorite for biological control in the early part of the treatment process. Hypochlorite is removed by the oxygen scavenger, which results in a slight increase in the required level of chemical injection.

10.2 Microbially Induced Corrosion. Although oxygen is normally the main corrosive medium observed in water-injection systems, the fact that bacterial growth is often an issue for these systems introduces the possibility that microbially induced corrosion could also be an important process. This type of corrosion requires the presence of an active biofilm, which implies that, provided the biological prevention and control philosophy is adequately maintained, this corrosion mechanism will not occur. It is thus a process initiated in association with sessile bacterial populations—free-flowing planktonic bacteria will not cause any such problems. When the bacteria begin to adhere to a surface, they tend to generate extracellular polymeric material below which the bacteria will grow. The local conditions within the biofilm could be significantly different from the conditions within the bulk of the pipe flow, which could promote conditions under which corrosion occurs. Additionally, because the biofilm will not be prevalent on the entire metal surface, this implies that the associated corrosion will be localized, and consequently, pitting of the metal can often be observed.

A large range of bacterial species can be present in a system and therefore be available as potential participants in microbially induced corrosion processes. However, the conditions within a water-injection system are, by and large, anoxic, and it is thought to be very likely that microbially induced corrosion processes in this setting are dominated by the activity of SRB (Enning and Garrelfs 2014). This view is strengthened by the observation that SRB tend to produce greater levels of corrosion than other bacterial groups, as observed in laboratory-based tests. Nevertheless, it has been noted that oxygen ingress into water-injection systems is always possible at locations where leaks occur. Such ingress can lead to the formation of highly corrosive sulfur species from the partial oxidation of dissolved H_2S or iron sulfide deposits at steel surfaces, which could further exacerbate corrosion problems where damage from SRB has already been induced.

There is no consensus regarding a single mechanism for microbially induced corrosion. Possible mechanisms for this type of corrosion include

- Cathodic depolarization by hydrogenase: This was the first mechanism proposed, in the 1930s, for the microbially induced corrosion of metals. It is based on the assumption that the rate-limiting step in corrosion is the dissociation of hydrogen from the cathodic site. SRB can consume hydrogen through the action of their hydrogenase enzymes and consequently "depolarize" the cathode, thus accelerating corrosion. However, the previously observed acceleration of cathodic reactions in SRB cultures could be explained by the reaction between sulfide and iron rather than by the microbial consumption of cathodic hydrogen. Furthermore, culture-based experiments have been unable to demonstrate that the bacterial consumption of cathodic hydrogen accelerates iron corrosion to any significant extent. As a result, this is now generally discounted as an important process in microbially induced corrosion.
- Underdeposit acid attack: Acetic acid or other acids could be among the final products of metabolic pathways in biofilm communities. This could induce

significant corrosion to the metal when the acid is concentrated within the confines of the biofilm. Some bacteria are also capable of generating sulfuric acid as a waste product that would be capable of generating very high corrosion rates.

- Formation of an occluded area on the metal surface: One of the reasons that biofilms are not formed at all locations in a pipeline is that local conditions are needed around which the biofilm can begin to form. Conditions that might promote colonization include the local roughness of the pipe wall, inclusions, the surface charge, or already existing corrosion sites. The local conditions in the biofilm are very different from those in the remainder of the pipe, and this could lead to the formation of crevices and ion concentration cells that results in accelerated local corrosion rates.
- Fixing anodic sites: Because the bacterial population is fixed at the site of the biofilm, the anodic site becomes "fixed," which leads to localized pitting. This is one reason that most cases of microbially induced corrosion are characterized by pitting-type corrosion.
- Electrical microbially induced corrosion (Venzlaff et al. 2013): Some strains of SRB have been shown to be capable of being directly fueled by the consumption of iron-derived electrons, without the involvement of cathodic hydrogen gas as an intermediate. This is known as electrical microbially induced corrosion.

Irrespective of the exact mechanism (and a number of different microbially induced corrosion mechanisms are probably occurring in any given application), it is evident that this process has the potential to increase corrosion in water-injection systems to higher levels than would otherwise be expected. In one corrosion study in a North Sea water-injection system, the observed corrosion rates as measured using corrosion coupons were an order of magnitude greater than those expected on the basis of the corrosivity of the abiotic, anoxic water (Comanescu et al. 2012) (**Fig. 39**). This significant increase in the corrosion rate was ascribed to microbially induced corrosion.

Fig. 39—Corrosion coupons before (left) and after (right) 6 months in a water-injection system (Comanescu et al. 2012).

10.3 Corrosion Control in Water-Injection Systems. Corrosion control in production systems is often focused on methods to passivate the metal surface using film-forming inhibitors. Such methodologies are rarely applied in water-injection systems because the destination for all fluids in this process is the injection well and concerns regarding the impacts on injectivity tend to steer operators away from these solutions.

Control in water-injection systems is therefore focused on prevention. In terms of the two main corrosion types described in this section, this means ensuring robust oxygen and biological control. Short excursions from the oxygen specification will increase the corrosion rates for the duration of the excursion, but this can often be managed, provided that the process upsets are not persistent.

The consequences of poor bacterial control can be more serious. After a biofilm has become established, the generation of biopolymers can limit the ability of biocides to penetrate the biofilm, which can render the biocides less effective. For biological control, prevention rather than cure is thus a vitally important approach. Unfortunately, it is one that seems to be rarely applied, and as a result, biological problems are frequently encountered. The consequences are wide reaching, and in addition to increased localized corrosion problems, this can induce exacerbated reservoir souring, reduced well injectivity, and possibly reduced sweep as a result of preferential plugging of lower-permeability intervals.

11. Water Sourcing for Waterfloods
For any given project, there is usually a range of potential water sources that might be used for waterflood purposes. Those water sources will have different chemistries and productivities, deliver different water qualities, and carry different costs for the project. The selection of the appropriate source, or mixture of sources, can thus have a material impact on project success.

11.1 Selection Factors. When a range of available water sources has been identified, compositional data, availability and deliverability, temperature, and water quality are factors of interest in helping to decide the best water source to use.

11.1.1 Compositional Data. The type and concentration of dissolved ions influence scaling, corrosion, and the risk of swelling reservoir clays. It is also important to quantify the concentrations of any dissolved gases present (oxygen, CO_2, H_2S) because these also impact scaling and corrosion risks. CO_2 will be in equilibrium with bicarbonate/carbonate and will thus strongly influence the carbonate scaling risk.

If the salinity of a water source is low, there could be potential recovery benefits associated with a low-salinity effect. However, if it is too low, it could reduce injectivity through the swelling of reservoir clays or through the mobilization of clay fines.

The most important components to analyze and quantify include

- Sodium: Sodium is often used to adjust the cation/anion balance in the water source when they are not found to be in balance in the analysis.
- Calcium: Calcium is used to calculate calcium carbonate and calcium sulfate scaling tendencies. The calcium content of oilfield waters can range from a few hundred to 50 000 mg/L. High concentrations of calcium tend to precipitate

treating chemicals such as water-soluble corrosion inhibitors, scale inhibitors, and oxygen scavengers.

- Magnesium: Magnesium is used to calculate scaling tendencies for calcium sulfate. Magnesium is also a factor in forming dolomite (a calcium magnesium carbonate mineral).
- Barium: Very small concentrations of barium can cause serious barium sulfate scaling problems because it has very low solubility. The barium content in water can range from very low (1 to 10 mg/L) to very high (greater than 1000 mg/L).
- Strontium: Strontium is very similar to barium in its chemical properties. Slightly higher strontium levels are needed to induce precipitation compared to barium.
- Iron: The iron content of water can be used as a tool to monitor corrosion in some situations. Iron can be present as ferric iron (Fe^{3+}) or ferrous iron (Fe^{2+}).
- Chloride: Chlorides are a measure of the salinity of waters—content can range from approximately 100 mg/L up to 300 000 mg/L (salt saturated). Chloride concentration is an important factor in calculating scaling tendencies for all oilfield scales.
- Bicarbonate: This is very important in determining scaling tendencies for calcium carbonate. Bicarbonates are unstable, so the analysis should be performed in the field to obtain accurate information for scaling tendency calculations. The values can range from 0 to 10 000 mg/L.
- Carbonate: Carbonates are rarely present in produced waters because they exist only when the pH is greater than 8.3.
- Sulfate: The sulfate concentration is used to calculate scaling tendencies for calcium sulfate, barium sulfate, and strontium sulfate.
- pH: This is an important parameter in determining calcium carbonate scaling tendencies and the corrosivity of water. This analysis must be performed in the field.
- H_2S: This analysis must also be performed in the field. Concentrations can range from 0 to well in excess of 1000 mg/L.
- CO_2: This analysis also must be performed in the field. CO_2 is very important in calculating calcium carbonate scaling tendencies. Dissolved CO_2 content typically ranges from 0 to 2000 mg/L. There is also a need to consider CO_2 analysis in the gas phase of a system.
- Oxygen: Dissolved oxygen contributes significantly to corrosion problems by enhancing the corrosion reactions and accelerating H_2S and CO_2 corrosion. Oxygen reacts with H_2S to form elemental sulfur, which accelerates corrosion and the plugging tendencies of the water. Oxygen facilitates the growth of aerobic bacteria. It is absent from formation waters in situ, but it can enter through the topside process.

11.1.2 Availability and Deliverability. It is important to understand the quantity of water needed for a project and if the prospective water source can deliver those volumes. There could be volumetric constraints associated with aquifers that are of a limited extent. Just as for an oil reservoir, it is important to understand the reservoir properties for aquifer sources—this will affect well deliverability and, hence, the number of water-supply wells needed to supply the required volumes.

It therefore has a clear impact on the cost of supplying the water. There might also be a need to consider if the reservoir properties have areal variations that might limit the deliverability in certain parts of the field.

11.1.3 Temperature. The injection-water temperature affects scale and corrosion risks and plays a very strong role in influencing the likelihood of inducing fractured injection and the expected injectivity. (The reasons for this are explained in detail in *Waterflooding: Injection Regime and Injection Wells*, another book in this series.) A relatively hot water source could confer reduced injectivity in water-injection wells. However, in paraffinic crudes, it could help to avoid wax dropout.

11.1.4 Water Quality. The suspended-solids loading has a big impact on the extent of impairment and injectivity decline under matrix injection. It also fundamentally influences induced fracture half-length under fractured injection. For water sources that might supply water with high solids loading, there could be a need for increased upfront investment in water-treatment equipment.

11.2 Water-Source Options. The potential water sources and their impacts are discussed in the sections that follow.

11.2.1 Seawater. For offshore projects, seawater is a commonly used water source because an abundant supply is available. The composition of seawater changes slightly depending on the region, although the changes are not usually such that they have an impact on the issues associated with seawater injection. There can often be local changes in composition, such as in locations close to river outflows, where the seawater is likely to have a lower salinity and perhaps an increase in the dispersed-solids composition.

There are also regional composition variations. These differences usually arise because of differences in the balance between surface evaporation and precipitation and the extent of mixing between surface waters and deeper water. The variations in composition that are possible highlight the importance of quantifying the composition of seawater at the location of all new waterflood developments because these differences can influence the chemistry risks and exposures—they can induce different levels of scaling problems, for example.

In addition, there could well be some compositional differences related to depth; hence, when seawater is being considered as a potential water source, there should be an analytical campaign to assess samples from different depths at the location to optimize the intake depth. The most important seawater property changes affected by depth are usually oxygen content, temperature, and (sometimes) suspended-solids content.

One of the key compositional components of seawater is dissolved oxygen, which will induce corrosion problems in a water-injection system if it is not removed. There are two principal sources of oxygen in seawater—diffusion across the air/sea interface and photosynthetic processes. These result in a variation in oxygen concentration with depth, as illustrated in **Fig. 40**. At the surface, the concentration is commonly present in the range of 5 to 8 mL/L. The oxygen-concentration minimum is observed in the pycnocline layer, which is the layer with the greatest density gradient. Organisms are drawn to this layer because of its ample food supply, and their respiration processes result in some depletion of the oxygen levels. The oxygen concentration at deeper levels is derived from surface layers, but the concentration increases because food is scarcer at deeper levels, so depletion from respiration processes is less.

Generally, oxygen-concentration changes because of depth do not act as a significant driver for the water-intake depth because the efficiency of mechanical oxygen-removal processes is such that there is no significant difference in the amount of chemical oxygen scavenger that is required.

Changes in water temperature can drive changes to the intake depth, as can suspended-solids loading because rivers—which are often heavily laden with sediments—can locally increase solids loading at shallow depths. The temperature profile for seawater is, not surprisingly, dependent on location—generic profiles are shown in **Fig. 41.** There can also be seasonal changes in temperature. For reservoirs where the formation-water temperature is very high, there could be a driver to increase the seawater-intake depth because the seawater is commonly also used for process cooling needs, and deepening could then reduce the cooling-water volume requirements.

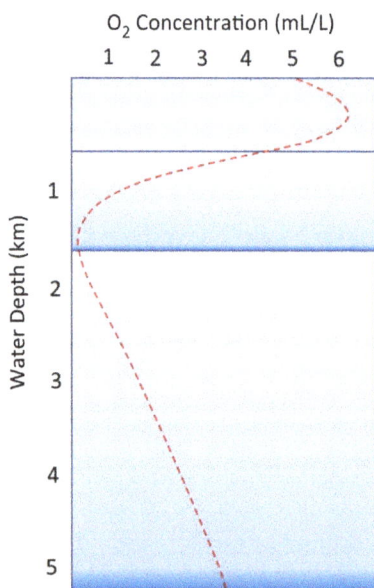

Fig. 40—Seawater oxygen-concentration relationship with depth.

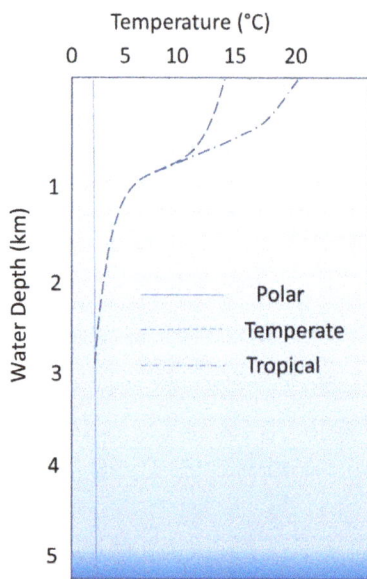

Fig. 41—Seawater temperature profile with depth.

One of the key issues associated with the use of seawater is the potential for scaling. Seawater is almost completely saturated with respect to calcium carbonate, but this does not typically induce significant problems. The primary scaling problems for seawater systems relate to the presence of appreciable concentrations of sulfate (usually in the range of 2400 to 3200 mg/L), which—depending on the concentrations of calcium, strontium, and barium in the formation water—could induce appreciable sulfate-scaling problems. When those scales manifest at the producer perforations, there will likely be a need for scale squeeze treatments or, in severe cases, a reduction of the sulfate content of the seawater. The presence of sulfate also means that seawater floods invariably induce reservoir souring, as previously described. With extreme scaling and souring problems, there might be a case to be made to use an aquifer water source instead of seawater.

Raw-seawater injection, using only a coarse water treatment, has been used in Statoil's Norne Field, in the Norwegian Sea, since 1997. This option offers the benefit of reduced facilities costs, but because oxygen is not removed, the deployment of high-grade corrosion-resistant materials is needed. This tends to render the option a niche application. However, a similar application at the Barton Field in Malaysia reportedly reduced the original engineering capital expenditure by 60% (Lee et al. 2004). In the Norne case, the operator also believes that the injection of aerated water stimulates aerobic bacteria and that there might also be an improved recovery benefit associated with interfacial-tension reduction.

An additional variation is to place the water-injection facilities on the seabed, reducing the topside load even further (Waite et al. 1996). This option has been used on Columba E (UK North Sea) (Rogerson and Laing 2007), Tyrihans (Norway) (Grynning et al. 2009), and Albacora (Brazil) (Buk et al. 2013). Each of these projects uses simple filtration, no electrochlorination, and no sulfate removal. Again, the limited number of applications suggests that this could be a niche technology.

11.2.2 Aquifer Sources. A number of potential aquifer sources could be available for any given location. In considering such water sources, it is important to not only establish the water quality to be expected from the reservoir (so that it can be decided if any water treatment is needed) but also assess the reservoir deliverability, in terms of both volume and sustainability, in much the same way that an oil reservoir is normally assessed. The deliverability will determine the number of water supply wells that will be needed. These considerations are vitally important, but the answers might not be readily available because, although the aquifers might be found during the drilling of oil wells, they are not normally tested for productivity and water quality.

One potential benefit of aquifer water sources is the opportunity to operate a dumpflood project. This option uses the water supply wells as injection wells, thereby reducing both surface infrastructure and well costs. A dumpflood is most easily achieved if there is a pressure differential between the water source and the target reservoir that allows the water to naturally crossflow into the target reservoir, although it is possible to install a supplementary pumping capability to accelerate injection rates. A potential drawback of the dumpflooding technique is that unless a relatively sophisticated monitoring and surveillance program is put in place, there is a danger that insufficient data are obtained regarding injection volumes and where those volumes go in the target reservoir.

One potentially important concern associated with an aquifer water supply arises in cases where the aquifer water is potable or could be easily and cheaply treated to make it potable. Such concerns become heightened in areas where potable water sources are already limited. When shallow, relatively fresh aquifer sources are used, special care must be taken to ensure that any clay-related problems are properly managed if the oil-producing reservoir is a sandstone. Furthermore, shallow water supplies often contain appreciable amounts of dissolved oxygen, which might need to be removed.

11.2.3 Produced Water. As production from a waterflooded field matures, it is inevitable that the volumes of produced water will increase. This water must be disposed of in some way, and because it contains suspended oil, there could be environmental concerns associated with its disposal. The disposal option that carries the lowest environmental impact appears to be to return the water to the reservoir from which it came. This option has the added benefit of reducing the volumes of supplemental makeup water required for the project.

In the Ula Field in Norway, Bakke et al. (1996) suggested that PWRI would reduce the discharge of chemicals and natural produced-water constituents by approximately 90%. Environmental discharge specifications appear to be tightening in many locations, which is resulting in increasing numbers of projects using produced water for injection purposes.

This increasingly important driver is balanced by a number of materially significant problems associated with PWRI. One obvious drawback is the volume of produced water available. Assuming that the field is not part of a larger cluster, if water injection is applied from the outset, produced water will not be initially available. Even when it does become available, a supplemental water source will usually be needed to replace the off-take of the hydrocarbons. Therefore, the use of produced water typically results in the use of at least two separate water sources for injection. Usually, those sources need to be mixed, increasing the potential for scaling problems. If the second water source contains sulfate (e.g., seawater), there is almost always a significantly heightened reservoir-souring risk because the seawater provides sulfate and the produced water is usually a good source of VFAs. (This was explained in detail in Section 3.3.)

Another problem associated with produced water is injectivity. Part of that problem is related to the fact that produced water is usually one of the dirtiest water sources available for use. Not only does it contain significant amounts of suspended solids (likely to be a combination of reservoir fines, corrosion products, and scales), it also contains suspended oil droplets, and the suspended oil often impairs synergistically with the suspended solids. Produced-water treatment is possible, of course, but because the source is very dirty to begin with, it is often the case that capital expenditure is significant and the result is still relatively dirty water.

The higher temperature commonly associated with produced water also has a negative impact on injectivity. Because produced water is relatively dirty, such projects typically use fractured injection, and the temperature dependence can be explained by the changing aperture of permeable fractures resulting from thermal expansion/contraction (Gunnarsson 2013), and cooler injection water tends to generate longer fractures that benefit injectivity.

Consequently, as a project using PWRI matures, it will gradually inject a greater proportion of produced water, which means that the injection-water temperature begins to rise. Consequently, a gradual decline in injectivity will be encountered, with associated negative impacts on pore-volume throughput or, alternatively, the need for additional water-injection wells. Martins et al. (1995) looked at injectivity changes seen in Prudhoe Bay as the waterflood was changed between seawater and produced-water-injection sources and concluded that although water quality and viscosity (this factor actually favors injectivity for warmer water) were both import-ant factors influencing injectivity, it was the thermal effects associated with PWRI that were responsible for the lower injectivity with that water source. Martins et al. (1995) also stated that injectivities with PWRI were 30 to 50% lower than those for seawater.

The lower injectivity associated with PWRI implies that there is a greater pumping power needed for a unit volume of injection. Vik (2008) reports on the calculated impact this had on gaseous emissions for an operator in Norway in 1999. The indi-cation was that PWRI increased discharges of CO_2 by 2.4 million tons and the dis-charge of nitrogen oxide gases by 11,000 tons (implying gaseous-discharge increases of between 40 and 80%). This suggests that the environmental impacts of PWRI might not be as evident as originally suggested. The findings from Vik (2008) imply that if any field were to use PWRI claiming a reduction in environmental impacts, there must have been an analysis that examined the relative environmental impacts of gaseous discharges vs. produced-water discharges. Processes are already in place that enable such analyses to be made (Bakke et al. 1996).

Vik (2008) also suggested that the injection uptime for PWRI might be lower than that for seawater injection.

One additional issue to be aware of in relation to PWRI is the potential for corro-sion. This might result from the presence of CO_2, but there is also a definite corro-sion risk related to the presence of oxygen. It is recognized that oxygen is not initially present when the produced water is present in the reservoir. This can actually cause a problem because this assumption is often retained throughout the topside pro-cess, but as has already been noted, leaks will very often introduce oxygen into produced-water systems. In such cases, there could be additional dangers involved because the tendency is to assume that oxygen is absent, and as a result, control is often absent.

Finally, the scaling risk for PWRI needs to be assessed. Many produced waters are saturated with calcium carbonate, which can be readily controlled by means of a scaling inhibition program. In extreme cases, uninhibited injection lines can plug completely with scale under shut-in conditions.

11.2.4 Other Water-Source Options. Although seawater, produced water, and aquifer water supplies are the three most commonly used water sources, local con-ditions could result in the use of alternatives. For example, Lake Maracaibo water is used in Venezuela, and Euphrates River water is used in Syria.

Even with major river sources, there could be strong competition for river-water usage, particularly in areas with only modest levels of rainfall, so particular atten-tion must be paid to ensure that the use of such water sources is demonstrably sustainable.

12. Conclusions

In many early waterflood projects, chemistry issues were not recognized as key focus areas. Indeed, some of the issues—such as reservoir souring and low-salinity flooding—had not yet been identified. In current waterfloods, the situation has sig-nificantly improved, although a comprehensive understanding of the full impact of chemistry issues, and a subsequent optimization program, is often still missing.

Chemistry issues have a significant impact on the operating cost to be borne by a waterflood development. This is often especially true if the full impact of the issues is not recognized at the outset and control has to be implemented on a remedial basis. Many of these issues will not only affect the operating cost but also can significantly affect the subsequent production and recovery profile. This does not apply only to low-salinity flooding, in which the chemistry is specifically manipulated to influence recovery. Recovery can also be negatively influenced by reservoir souring if, for example, the produced H_2S levels reach the point at which producers have to be shut in because materials limits have been reached.

It is evident, therefore, that chemistry aspects need to be robustly addressed in modern waterflood developments to reduce expensive remediation actions, optimize operating costs, facilitate production optimization, and maximize recovery.

13. Nomenclature

$Al_2Si_2O_5(OH)_4$	=	kaolinite
Al	=	aluminum
Ba	=	barium
B/D	=	barrels per day
BSW	=	basic sediment and water
BTX	=	benzene, toluene, and xylene
BWPD	=	barrels of water per day
C	=	carbon
Ca	=	calcium
C_6H_6	=	benzene
$C_6H_4(CH_3)_2$	=	xylene
$C_6H_5CH_3$	=	toluene
CH_3COO^-	=	acetate ion
CH_3OH or MeOH	=	methanol
Cl^-	=	chloride ion
ClO_2^-	=	chlorite
ClO_3^-	=	chlorate
cm^3/min	=	cubic centimeters per minute
cm^3/mol	=	cubic centimeters per mole
CO_2	=	carbon dioxide
CO_3^{2-}	=	carbonate ion
COO^-	=	carboxylate ion
Cr	=	chromium
Cs	=	cesium
DTPA	=	diethylenetriaminepentaacetic acid
DVA	=	direct vertical access
e^-	=	electron
EDTA	=	ethylenediaminetetraacetic acid
EOR	=	enhanced oil recovery
ESP	=	electrical submersible pump
Fe	=	iron
ft	=	foot
g	=	gram
g/L	=	grams per liter

H	=	hydrogen
H$^\bullet$	=	unassociated hydrogen atom
HCO_3^-	=	bicarbonate ion
H_2O	=	water
HP	=	high pressure
HS^-	=	bisulfide ion
H_2S	=	hydrogen sulfide
HSE	=	health, safety, and environment
HSO_3^-	=	bisulfite ion
K	=	potassium
kbd	=	thousands of barrels per day
kg	=	kilogram
kg/d	=	kilograms per day
$kJ \cdot mol^{-1}$	=	kilojoules per mole
km	=	kilometer
KOH	=	potassium hydroxide
lbm	=	pounds mass
Li	=	lithium
LP	=	low pressure
m	=	meter
m^3	=	cubic meter
md	=	millidarcy
m^3/d	=	cubic meters per day
MeOH or CH_3OH	=	methanol
MEOR	=	microbial enhanced oil recovery
meq/100 g	=	milliequivalents per 100 grams
meq/L	=	milliequivalents per liter
mg	=	milligram
m^2/g	=	square meters per gram
Mg	=	magnesium
mg/L	=	milligrams per liter
mg/m^2	=	milligrams per square meter
MHGC	=	Medicine Hat Glauconitic C
MIC	=	minimum inhibitor concentration
mL/L	=	milliliters per liter
mm/yr	=	millimeters per year
m/s	=	meters per second
N	=	nitrogen
Na	=	sodium
NO_3^-	=	nitrate ion
NRB	=	nitrate-reducing bacteria
NTMP	=	nitrilotri(methylenephosphonic) acid
O	=	oxygen
OH^-	=	hydroxyl ion
OHL	=	openhole log
OOIP	=	original oil in place
P	=	phosphorus
Phn	=	phosphonate

pKa	=	negative log of acid dissociation constant
ppb	=	parts per billion
ppm	=	parts per million
ppmv	=	parts per million by volume
PRB	=	perchlorate-reducing bacteria
psi	=	pounds per square inch
PWRI	=	produced-water reinjection
Ra	=	radium
Rb	=	rubidium
RFT	=	repeat formation test
ROS	=	residual oil saturation
s	=	solid
S	=	sulfur
SCAL	=	special core analysis
SO_3^{2-}	=	sulfite ion
SO_4^{2-}	=	sulfate ion
SONRB	=	sulfide-oxidizing nitrate-reducing bacteria
Sr	=	strontium
SRB	=	sulfate-reducing bacteria
SRU	=	sulfate-removal unit
STB/D	=	stock-tank barrels per day
S_w	=	water saturation
SWI	=	seawater injection
TDS	=	total dissolved solids
Th	=	thorium
THPS	=	tetrakis(hydroxymethyl)phosphonium sulfate
TVS	=	thermal viability shell
U	=	uranium
UK	=	United Kingdom
US	=	United States
USD	=	US dollar
VFA	=	volatile fatty acid
WCT	=	water cut
WF	=	waterflood
w/v	=	weight per volume
w/w	=	weight per weight
x_D	=	dimensionless distance
$\Delta G^{o\prime}$	=	Gibbs free energy (or free enthalpy)
µm/yr	=	microns per year

14. References

Al Harrasi, A., Al-Maamari, R. S., and Masalmeh, S. K. 2012. Laboratory Investigation of Low Salinity Waterflooding for Carbonate Reservoirs. Paper presented at the Abu Dhabi International Petroleum Conference and Exhibition, Abu Dhabi, UAE, 11–14 November. SPE-161468-MS. https://doi.org/10.2118/161468-MS.

Alkindi, A., Prince-Wright, R., Moore, W. R. et al. 2008. Challenges for Waterflooding in a Deepwater Environment. *SPE Prod & Oper* **23** (3): 404–410. SPE-118735-PA. https://doi.org/10.2118/118735-PA.

Al-Mayahi, N. M., Snippe, J., Rucci, F. D. et al. 2012. Water Injection Subsurface Challenges and Reactive Transport Modelling. Paper presented at the SPE EOR Conference at Oil and Gas West Asia, Muscat, Oman, 16–18 April. SPE-154457-MS. https://doi.org/10.2118/154457-MS.

Al-Rasheedi, S., Kalli, C., Thrasher, D. et al. 1999. Prediction and Evaluation of the Impact of Reservoir Souring in North Kuwait, A Case Study. Paper presented at the Middle East Oil Show and Conference, Bahrain, 20–23 February. SPE-53164-MS. https://doi.org/10.2118/53164-MS.

Al-Refai, S. A., Al-Ajmi, M., Oduola, L. et al. 2019. Souring Prediction on Mature Waterflooded Reservoirs in North Kuwait. Paper presented at SPE Europec featured at 81st EAGE Conference and Exhibition, London, England, UK, 3–6 June. SPE-195561-MS. https://doi.org/10.2118/195561-MS.

Al-Riyami, M. M., Mackay, E. J., Deliu, G. et al. 2008. When Will Low-Sulphate Seawater No Longer Be Required on the Tiffany Field? Paper presented at the SPE International Symposium and Exhibition on Formation Damage Control, Lafayette, Louisiana, USA, 13–15 February. SPE-112537-MS. https://doi.org/10.2118/112537-MS.

Arensdorf, J. J., Miner, K., Ertmoed, R. et al. 2009. Mitigation of Reservoir Souring by Nitrate in a Produced-Water Reinjection System in Alberta. Paper presented at the SPE International Symposium on Oilfield Chemistry, The Woodlands. Texas, USA, 20–22 April. SPE-121731-MS. https://doi.org/10.2118/121731-MS.

ASTM D1141-98, Standard Practice for the Preparation of Substitute Ocean Water. 2003. West Conshohocken, Pennsylvania: ASTM International. https://doi.org/10.1520/D1141-98R13.

Austad, T., Strand, S., Høgnesen, E. J. et al. 2005. Seawater as IOR Fluid in Fractured Chalk. Paper presented at the SPE International Symposium on Oilfield Chemistry, The Woodlands, Texas, USA, 2–4 February. SPE-93000-MS. https://doi.org/10.2118/93000-MS.

Ayirala, S. C., Uehara-Nagamine, E., Matzakos, A. N. et al. 2010. A Designer Water Process for Offshore Low Salinity and Polymer Flooding Applications. Paper presented at the SPE Improved Oil Recovery Symposium, Tulsa, Oklahoma, USA, 24–28 April. SPE-129926-MS. https://doi.org/10.2118/129926-MS.

Badrak, R.P. 2018. NACE MR0175/ISO 15156: Update on Current Document and Where Are We Going? Paper presented at CORROSION 2018, Phoenix, Arizona, USA, 15–19 April. NACE-2018-11294.

Bakke, S., Ofjord, G. D., Vik, E. A. et al. 1996. Environmental Risk Management of Produced Water – A Demonstration of the CHARM Model Used for the Ula Field in the North Sea. Paper presented at the SPE Health, Safety and Environment in Oil and Gas Exploration and Production Conference, New Orleans, Louisiana, USA, 9–12 June. SPE-35821-MS. https://doi.org/10.2118/35821-MS.

Bernard, G. G. 1967. Effect of Floodwater Salinity on Recovery of Oil from Cores Containing Clays. Paper presented at the SPE California Regional Meeting, Los Angeles, California, USA, 26–27 October. SPE-1725-MS. https://doi.org/10.2118/1725-MS.

Billo, S.M. 1987. Petrology and Kinetics of Gypsum-Anhydrite Transitions. Journal of Petroleum Geology 10 (1): 73–86. https://doi.org/10.1111/j.1747-5457.1987.tb00997.x.

Bittner, S. D., Zemlak, K. R., and Korotash, B. D. 2000. Coiled Tubing Scale Removal of Iron Sulfide – A Case Study of the Kaybob Field in Central Alberta. Paper presented at the SPE/ICoTA Coiled Tubing Roundtable, Houston, Texas, USA, 5–6 April. SPE-60695-MS. https://doi.org/10.2118/60695-MS.

Bogaert, P., Berredo, M. C., Toschi, C. et al. 2006. Scale Inhibitor Squeeze Treatments Deployed from an FPSO in a Deepwater Subsea Fields in the Campos Basin. Paper presented at the SPE Annual Technical Conference and Exhibition, San Antonio, Texas, USA, 24–27 September. SPE-102503-MS. https://doi.org/10.2118/102503-MS.

Brown, A. D. F., Merrett, S. J., and Putnam, J. S. 1991. Coil-Tubing Milling/Underreaming of Barium Sulphate Scale and Scale Control in the Forties Field. Paper presented at Offshore Europe, Aberdeen, UK, 3–6 September. SPE-23106-MS. https://doi.org/10.2118/23106-MS.

Bryant, S. L. and Lockhart, T. P. 2000. Reservoir Engineering Analysis of Microbial Enhanced Oil Recovery. Paper presented at the SPE Annual Technical Conference and Exhibition, Dallas, Texas, USA, 1–4 October. SPE-63229-MS. https://doi.org/10.2118/63229-MS.

Buk, L., Jr., Costa, O. C., de Siqueira, A. G. et al. 2013. Albacora Subsea Raw Water Injection Systems. Paper presented at the Offshore Technology Conference, Houston, Texas, USA, 6–9 May. OTC-24167-MS. https://doi.org/10.4043/24167-MS.

Burger, E. D. and Jenneman, G. E. 2009. Forecasting the Effects of Reservoir Souring from Waterflooding a Formation Containing Siderite. Paper presented at the SPE International Symposium on Oilfield Chemistry, The Woodlands, Texas, USA, 20–22 April. SPE-121432-MS. https://doi.org/10.2118/121432-MS.

Burger, E. D., Jenneman, G. E., and Gao, X. 2013. The Impact of Dissolved Organic-Carbon Type on the Extent of Reservoir Souring. Paper presented at the SPE International Symposium on Oilfield Chemistry, The Woodlands, Texas, USA, 8–10 April. SPE-164068-MS. https://doi.org/10.2118/164068-MS.

Burger, E. D., Jenneman, G. E., Vedvik, A. et al. 2006. Forecasting the Effect of Produced Water Reinjection on Reservoir Souring in the Ekofisk Field. Paper presented at CORROSION 2006, San Diego, California, USA, 12–16 March. NACE-06661.

Burger, E. D and Odom, J. M. 1999. Mechanisms of Anthraquinone Inhibition of Sulphate-Reducing Bacteria. Paper presented at the SPE International Symposium on Oilfield Chemistry, Houston, Texas, USA, 16–19 February. SPE-50764-MS. https://doi.org/10.2118/50764-MS.

Burger, E., Venkat, P. S., and Mittal, S. 2019. Modeling/Forecasting Reservoir Souring in a Field Rajasthan, India with an Extremely Low Indigenous Volatile Fatty Acid VFA Concentration. Paper presented at the SPE International Conference on Oilfield Chemistry, Galveston, Texas, USA, 8–9 April. SPE-193636-MS. https://doi.org/10.2118/193636-MS.

Cassinat, J. C., Payette, M. C., Taylor, D. B. et al. 2002. Optimizing Waterflood Performance by Utilizing Hot Water Injection in a High Paraffin Content Reservoir. Paper presented at the SPE/DOE Improved Oil Recovery Symposium, Tulsa, Oklahoma, USA, 13–17 April. SPE-75141-MS. https://doi.org/10.2118/75141-MS.

Cavallaro, A. N., Martinez, M. E. G., Ostera, H. et al. 2005. Oilfield Reservoir Souring During Waterflooding: A Case Study with Low Sulphate Concentration in Formation and Injection Waters. Paper presented at the SPE International

Symposium on Oilfield Chemistry, The Woodlands, Texas, USA, 2–4 February. SPE-92959-MS. https://doi.org/10.2118/92959-MS.

Clarke, T. A. and Nasr-El-Din, H. A. 2015. Application of a Novel Clay Stabilizer to Mitigate Formation Damage Due to Clay Swelling. Paper presented at the SPE Production and Operations Symposium, Oklahoma City, Oklahoma, USA, 1–5 March. SPE-173625. https://doi.org/10.2118/173625-MS.

Collins, I. R., Jordan, M. M., Feasey, N. et al. 2001. The Development of Emulsion-Based Production Chemical Deployment Systems. Paper presented at the SPE International Symposium on Oilfield Chemistry, Houston, Texas, USA, 13–16 February. SPE-65026-MS. https://doi.org/10.2118/65026-MS.

Collins, S. H. and Melrose, J. C. 1983. Adsorption of Asphaltenes and Water on Reservoir Rock Minerals. Paper presented at the SPE Oilfield and Geothermal Chemistry Symposium, Denver, Colorado, USA, 1–3 June. SPE-190385-PA. https://doi.org/10.2118/11800-MS.

Comanescu, I., Taxen, C., and Melchers, R. E. 2012. Assessment of MIC in Carbon Steel Water Injection Pipelines. Paper presented at the SPE International Conference & Workshop on Oilfield Corrosion, Aberdeen, UK, 28–29 May. SPE-155199-MS. https://doi.org/10.2118/155199-MS.

Cord-Ruwisch, R., Kleinitz, W., and Widdel, F. 1987. Sulfate-Reducing Bacteria and Their Activities in Oil Production. *J Pet Technol* 39 (1): 97–106. SPE-13554-PA. https://doi.org/10.2118/13554-PA.

Cusack, F., McKinley, V. L., Lappin-Scott, H. M. et al. 1987. Diagnosis and Removal of Microbial/Fines Plugging in Water Injection Wells. Paper presented at the SPE Annual Technical Conference and Exhibition, Dallas, Texas, USA, 27–30 September. SPE-16907-MS. https://doi.org/10.2118/16907-MS.

de Jesus, E. B., de Andrade Lima, L. R. P, Bernardez, L. A. et al. 2015. Inhibition of Microbial Sulfate Reduction by Molybdate. *Brazilian Journal of Petroleum and Gas* 9 (3): 95–106. https://doi.org/10.5419/bjpg2015-0010.

Delshad, M., Bryant, S. L., Sepehrnoori, K. et al. 2009. Development of a Reservoir Simulator for Souring Predictions. Paper presented at the SPE Reservoir Simulation Symposium, The Woodlands, Texas, USA, 2–4 February. SPE-118951-MS. https://doi.org/10.2118/118951-MS.

De Vries, H. and Arnaud, F. 1993. The First Successful Downhole Barium Sulphate Scale Dissolving Operation. Paper presented at Offshore Europe, Aberdeen, UK, 7–10 September. SPE-26704-MS. https://doi.org/10.2118/26704-MS.

Dinning, A. J. 2011. A Review of PWRI Experience: Souring Mitigation, MIC Control and Monitoring. Presented at the TEKNA Produced Water Management Conference 2011, Stavanger, Norway.

Dolfing, J. and Hubert, C. R. J. 2017. Using Thermodynamics to Predict the Outcomes of Nitrate-Based Oil Reservoir Souring Control Interventions. *Frontiers in Microbiology* 8: 2575. https://doi.org/10.3389/fmicb.2017.02575.

Donaldson, J. D. and Grimes, S. M. 1987. Control of Scale in Sea Water Applications by Magnetic Treatment of Fluids. Paper presented at Offshore Europe, Aberdeen, UK, 8–11 September. SPE-16540-MS. https://doi.org/10.2118/16540-MS.

Dunsmore, B. C. and Evans. P. 2006. Reservoir Simulation of Sulfate-Reducing Bacteria Activity in the Deep Sub-Surface. Paper presented at CORROSION 2006, San Diego, California, USA, 12–16 March. NACE-06664.

Eden, B., Laycock, P. J., and Fielder, M. 1993. *Oilfield Reservoir Souring*. Health and Safety Executive – Offshore Technology Report, OTH 92 385.

Enerstvedt, E. and Boge, H. 2001. Scale Removal by Milling and Jetting with Coiled Tubing. Paper presented at the SPE/ICoTA Coiled Tubing Roundtable, Houston, Texas, USA, 7–8 March. SPE-68366-MS. https://doi.org/10.2118/68366-MS.

Enning, D. and Garrelfs, J. 2014. Corrosion of Iron by Sulfate-Reducing Bacteria: New Views of an Old Problem. *Applied and Environmental Microbiology* 80 (4): 1226–1236. https://doi.org/10.1128/aem.02848-13.

Evans, P., Nederlof, E., and Richmond, W. 2015. Souring Development Associated with PWRI in a North Sea Field. Paper presented at the SPE Produced Water Handling & Management Symposium, Galveston, Texas, USA, 20–21 May. SPE-174529-MS. https://doi.org/10.2118/174529-MS.

Feasey, N. D., Jordan, M. M., Mackay, E. J. et al. 2004. The Challenge That Completion Types Present to Scale Inhibitor Squeeze Chemical Placement: A Novel Solution Using a Self-Diverting Scale Inhibitor Squeeze Process. Paper presented at the SPE International Symposium and Exhibition on Formation Damage Control, Lafayette, Louisiana, USA, 18–20 February. SPE-86478-MS. https://doi.org/10.2118/86478-MS.

Fernandes, R. J. 1956. Sea Water for Water Flooding (Pacific Coast District Study Committee on Fluid Injection). In *Drilling and Production Practice*, API-56-185. New York: American Petroleum Institute.

Ferno, M. A., Gronsdal, R., Aheim, J. et al. 2011. Use of Sulfate for Water Based Enhanced Oil Recovery During Spontaneous Imbibition in Chalk. *Energy Fuels* 25 (4): 1697–1706. https://doi.org/10.1021/ef200136w.

Fischer, K. P., Salama, M. M., and Murali, J. 1996. Optimal Selection of Materials for Seawater Injection Systems: Testing in Deoxygenated Seawater. Paper presented at CORROSION 96, Denver, Colorado, USA, 24–29 March. NACE-96593.

Fjelde, I., Asen, S. M., Omekeh, A. et al. 2013. Secondary and Tertiary Low Salinity Water Floods: Experiments and Modeling. Paper presented at the EAGE Annual Conference & Exhibition incorporating SPE Europec, London, UK, 10–13 June. SPE-164920-MS. https://doi.org/10.2118/164920-MS.

Freitas, A. M., Lobão, A. C., and Cardoso, C. B. 2002. Hydrate Blockages in Flowlines and Subsea Equipment in Campos Basin. Paper presented at the Offshore Technology Conference, Houston, Texas, USA, 6–9 May. OTC-14257-MS. https://doi.org/10.4043/14257-MS.

Frenier, W. W. and Ziauddin, M. 2008. *Formation, Removal, and Inhibition of Inorganic Scale in the Oilfield Environment*. Richardson, Texas: Society of Petroleum Engineers.

Grynning, A., Larsen, S. V., and Skaale, I. 2009. Tyrihans Subsea Raw Seawater Injection System. Paper presented at the Offshore Technology Conference, Houston, Texas, USA, 4–7 May. OTC-20078-MS. https://doi.org/10.4043/20078-MS.

Gunnarsson, G. 2013. Temperature Dependent Injectivity and Induced Seismicity— Managing Reinjection in the Hellisheiði Field, SW-Iceland. *GRC Transactions* 37: 2013.

Gupta, R., Smith, G. G., Hu, L. et al. 2011. Enhanced Waterflood for Carbonate Reservoirs – Impact of Injection Water Composition. Paper presented at the SPE Middle East Oil and Gas Show and Conference, Manama, Bahrain, 25–28 September. SPE-142668-MS. https://doi.org/10.2118/142668-MS.

Haghshenas, M., Sepehrnoori, K., Bryant, S. L. et al. 2012. Modeling and Simulation of Nitrate Injection for Reservoir-Souring Remediation. *SPE J.* 17 (3): 817–827. SPE-141590-PA. https://doi.org/10.2118/141590-PA.

Halvorsen, E. N., Halvorsen, A. M. K., Andersen, T. R. et al. 2009. Qualification of Scale Inhibitors for Subsea Tiebacks with MEG Injection. Paper presented at the SPE International Symposium on Oilfield Chemistry, The Woodlands. Texas, USA, 20–22 April. SPE-121665-MS. https://doi.org/10.2118/121665-MS.

Heath, S., Selle, O. M., Storås, E. et al. 2014. Squeezing Sub-Sea Wells Co-Mingled in the Same Flowline on the Norne Field. Paper presented at the SPE International Oilfield Scale Conference and Exhibition, Aberdeen, Scotland, 14–15 May. SPE-169793-MS. https://doi.org/10.2118/169793-MS.

Heatherly, M. W., Howell, M. E., and McElhiney, J. E. 1994. Sulfate Removal Technology for Seawater Waterflood Injection. Paper presented at the Offshore Technology Conference, Houston, Texas, USA, 2–5 May. OTC-7593-MS. https://doi.org/10.4043/7593-MS.

Herbert, B. N., Gilbert, P. D., Stockdale, H. et al. 1985. Factors Controlling the Activity of Sulfate-Reducing Bacteria in Reservoirs During Water Injection. SPE-13978-MS.

Hitzman, D. O. 1994. A New Microbial Technique for Enhanced Oil Recovery and Sulphide Prevention and Reduction. Paper presented at the SPE/DOE Improved Oil Recovery Symposium, Tulsa, Oklahoma, USA, 17–20 April. SPE-27752-MS. https://doi.org/10.2118/27752-MS.

Horaska, D. D., Penkala, J. E., Reed, C. A. et al. 2009. Field Experiences Detailing Acrolein (2-propenal) Treatment of a Produced Water Injection System in the Sultanate of Oman. Paper presented at the SPE Middle East Oil and Gas Show and Conference, Manama, Bahrain, 15–18 March. SPE-120238-MS. https://doi.org/10.2118/120238-MS.

Hosseininoosheri, P., Lashgari, H., and Sepehrnoori, K. 2017. Numerical Prediction of Reservoir Souring Under the Effect of Temperature, Ph, and Salinity on the Kinetics of Sulfate-Reducing Bacteria. Paper presented at the SPE International Conference on Oilfield Chemistry, Montgomery, Texas, USA, 3–5 April. SPE-184562-MS. https://doi.org/10.2118/184562-MS.

Houston, S. J., Yardley, B., Smalley, P. C. et al. 2006. Precipitation and Dissolution of Minerals During a Waterflood – The Evidence of Produced Water Chemistry from Miller. Paper presented at the SPE International Oilfield Scale Symposium, Aberdeen, UK, 31 May–1 June. SPE-100603-MS. https://doi.org/10.2118/100603-MS.

Hu, Y. and Mackay, E. 2018. Reactive Transport Modelling of a Carbonate Reservoir Under Seawater Injection. Paper presented at the SPE International Oilfield Scale Conference and Exhibition, Aberdeen, Scotland, UK, 20–21 June. SPE-190757-MS. https://doi.org/10.2118/190757-MS.

Jack, T. R., Grigoryan, A., Lambo, A. et al. 2009. Troubleshooting Nitrate Field Injections for Control of Reservoir Souring. Paper presented at the SPE International Symposium on Oilfield Chemistry, The Woodlands, Texas, USA, 20–22 April. SPE-121573-MS. https://doi.org/10.2118/121573-MS.

Jack, T. R., Lee, E., and Mueller, J. 1984. Microbes and Oil Recovery. Proc., 2nd International Conference on Microbial Enhanced Oil Recovery, 167–180.

Jafar Fathi, S., Austad, T., and Strand, S. 2010. "Smart Water" as a Wettability Modifier in Chalk: The Effect of Salinity and Ionic Composition. Energy Fuels 24: 2514–2519. https://doi.org/10.1021/ef901304m.

James, J. S., Frigo, D. M., Heath, S. M. et al. 2005. Application of a Fully Viscosified Scale Squeeze for Improved Placement in Horizontal Wells. Paper presented at the

SPE International Symposium on Oilfield Scale, Aberdeen, UK, 11–12 May. SPE-94593-MS. https://doi.org/10.2118/94593-MS.

Jarrahian, K., Sorbie, K. S., Singleton, M. A. et al. 2019. The Effect of pH and Mineralogy on the Retention of Polymeric Scale Inhibitors on Carbonate Rocks for Application in Squeeze Treatments. *SPE Prod & Oper* **34** (2): 344–360. SPE-189519-PA. https://doi.org/10.2118/189519-PA.

Jenneman, G. E., Knapp, R. M., McInerney, M. J. et al. 1984. Experimental Studies of In-Situ Microbial Enhanced Oil Recovery. *SPE J.* **24** (1): 33–37. SPE-10789-PA. https://doi.org/10.2118/10789-PA.

Jenneman, G. E., McInerney, M. J., and Knapp, R. M. 1986. Effect of Nitrate on Biogenic Sulphide Production. *Applied and Environmental Microbiology* **51** (6): 1205–1211.

Jenneman, G. E., Moffitt, P. D., Baja, G. A. et al. 1997. Field Demonstration of Sulphide Removal in Reservoir Brine by Bacteria Indigenous to a Canadian Reservoir. Paper presented at the SPE Annual Technical Conference and Exhibition, San Antonio, Texas, USA, 5–8 October. SPE-38768-MS. https://doi.org/10.2118/38768-MS.

Jerauld, G. R., Webb, K. J., Lin, C.-Y. et al. 2008. Modeling Low-Salinity Waterflooding. *SPE Res Eval & Eng* **11** (6): 1000–1012. SPE-102239-PA. https://doi.org/10.2118/102239-PA.

Johnson, A., Eslinger, D., and Larsen, H. 1998. An Abrasive Jetting Scale Removal System. Paper presented at the SPE/ICoTA Coiled Tubing Roundtable, Houston, Texas, USA, 15–16 March. SPE-46026-MS. https://doi.org/10.2118/46026-MS.

Jones, A. M., Geissler, B., and De Paula, R. 2018. A Comparison of Chemistries Intended to Treat Reservoir Souring. Paper presented at CORROSION 2018, Phoenix, Arizona, USA, 15–19 April. NACE-2018-10583.

Jones, C., Downward, B., Edmunds, S. et al. 2012. THPS: A Review of the First 25 Years, Lessons Learned, Value Created and Visions for the Future. Paper presented at CORROSION 2012, Salt Lake City, Utah, USA, 11–15 March. NACE-2012-1505.

Jordan, M. 2019. The Impact of Production Logging Tool Data PLT on Scale Squeeze Lifetime and Management of Scale Risk in Norwegian Subsea Production Wells – A Case Study. Paper presented at the SPE Norway One Day Seminar, Bergen, Norway, 14 May. SPE-195618-MS. https://doi.org/10.2118/195618-MS.

Jordan, M. M., Collins, I. R., and Mackay, E. J. 2006. Low Sulfate Seawater Injection for Barium Sulfate Scale Control: A Life-of-Field Solution to a Complex Challenge. Paper presented at the SPE International Symposium and Exhibition on Formation Damage Control, Lafayette, Louisiana, USA, 15–17 February. SPE-98096-MS. https://doi.org/10.2118/98096-MS.

Jordan, M. M., Graham, G. M., Sorbie, K. S. et al. 1998. Scale Dissolver Application: Production Enhancement and Formation Damage Potential. Paper presented at the SPE Formation Damage Control Conference, Lafayette, Louisiana, USA, 18–19 February. SPE-39449-MS. https://doi.org/10.2118/39449-MS.

Jordan, M. M. and Mackay, E. J. 2005. Integrated Field Development for Effective Scale Control Throughout the Water Cycle in Deep Water Subsea Fields. SPE Europec/EAGE Annual Conference, Madrid, Spain, 13–16 June. SPE-94052-MS. https://doi.org/10.2118/94052-MS.

Jordan, M. M. and Mackay, E. J. 2009. The Challenge of Modelling and Deploying Diversion for Subsea Scale Squeeze Application. Paper presented at the

8th European Formation Damage Conference, Scheveningen, The Netherlands, 27–29 May. SPE-121376-MS. https://doi.org/10.2118/121376-MS.

Jordan, M. M., Murray, F., Kelly, A. et al. 2003. Deployment of Emulsified Scale – Inhibitor Squeeze to Control Sulphate/Carbonate Scales Within Subsea Facilities in the North Sea Basin. Paper presented at the International Symposium on Oilfield Chemistry, Houston, Texas, USA, 5–7 February. SPE-80249-MS. https://doi.org/10.2118/80249-MS.

Jordan, M. M., Sjuraether, K., Collins, I. R. et al. 2001. Life Cycle Management of Scale Control Within Subsea Fields and Its Impact on Flow Assurance, Gulf of Mexico and the North Sea Basin. Paper presented at the SPE Annual Technical Conference and Exhibition, New Orleans, Louisiana, USA, 30 September–3 October. SPE-71557-MS. https://doi.org/10.2118/71557-MS.

Jordan, M. M., Sorbie, K. S., Griffin, P. et al. 1995. Scale Inhibitor Adsorption/Desorption vs. Precipitation: The Potential for Extending Squeeze Life While Minimising Formation Damage. Paper presented at the SPE European Formation Damage Conference, The Hague, The Netherlands, 15–16 May. SPE-30106-MS. https://doi.org/10.2118/30106-MS.

Jordan, M. M., Sorbie, K. S., Ping, J. et al. 1994. Phosphonate Scale Inhibitor Adsorption/Desorption and the Potential for Formation Damage in Reconditioned Field Core. Paper presented at the SPE Formation Damage Control Symposium, Lafayette, Louisiana, USA, 7–10 February. SPE-27389-MS. https://doi.org/10.2118/27389-MS.

Kelland, M. A. 2009. Scale Control. In *Production Chemicals for the Oil and Gas Industry*, M. A. Kelland, Chap. 3, 53–110. Boca Raton, Florida: CRC Press.

Kelly, M., James, J. S., Frigo, D. M. et al. 2005. Application of Scale Dissolver and Inhibitor Squeeze Through the Gas-Lift Line in a Sub-Sea Field. Paper presented at the SPE International Symposium on Oilfield Scale, Aberdeen, UK, 11–12 May. SPE-95100-MS. https://doi.org/10.2118/95100-MS.

Kissel, C. L., Brady, J. L., Gottry, H. N. C. et al. 1985. Factors Contributing to the Ability of Acrolein to Scavenge Corrosive Hydrogen Sulfide. *SPE J.* 25 (5): 647–655. SPE-11749-PA. https://doi.org/10.2118/11749-PA.

Klepaker, J., Andrews, J., Vikane, O. et al. 2002. Successful Scale Stimulation Treatment of a Subsea Openhole Gravel-Packed Well in the North Sea. Paper presented at the International Symposium and Exhibition on Formation Damage Control, Lafayette, Louisiana, USA, 20–21 February. SPE-73715-MS. https://doi.org/10.2118/73715-MS.

Krishnan, C., Kopperson, D., and Cuthill, T. 1994. Discovery of Radioactive Barium Sulphate Scale in PanCanadian Petroleum Producing Operations in Southeastern Alberta. *J Can Pet Technol* 33 (10). PETSOC-94-10-06. https://doi.org/10.2118/94-10-06.

Kuijvenhoven, C., Noirot, J.-C., Bostock, A. M. et al. 2006. Use of Nitrate to Mitigate Reservoir Souring in Bonga Deepwater Development Offshore Nigeria. *SPE Prod & Oper* 21 (4): 467–474. SPE-92795-PA. https://doi.org/10.2118/92795-PA.

Lager, A., Webb, K. J., and Black, C. J. J. 2007. Impact of Brine Chemistry on Oil Recovery. Paper A24 presented at the 14th European Symposium on Improved Oil Recovery, Cairo, Egypt, 22–24 April.

Lager, A., Webb, K. J., Black, C. J. J. et al. 2006. Low Salinity Oil Recovery – An Experimental Investigation. Paper presented at the International Symposium of the Society of Core Analysts, Trondheim, Norway, 12–16 September.

Larsen, J., Sanders, P. F., and Talbot, R. E. 2000. Experience with the Use of Tetrak-ishydroxymethylphosphonium Sulphate (THPS) for the Control of Downhole Hydrogen Sulphide. Paper presented at CORROSION 2000, Orlando, Florida, USA, 26–31 March. NACE-00123.

Larsen, J., Rod, M. H., and Zwolle, S. 2004. Prevention of Reservoir Souring in the Halfdan Field by Nitrate Injection. Paper presented at CORROSION 2004, New Orleans, Louisiana, USA, 28 March–1 April. NACE-04761.

Lee, S. K.-H., Tan, T.-G., and Vaughan, A. 2004. An Attractive Raw Deal – Novel Approach to Barton Oilfield Water Injection. Paper presented at the SPE Asia Pacific Conference on Integrated Modelling for Asset Management, Kuala Lumpur, Malaysia, 29–30 March. SPE-87000-MS. https://doi.org/10.2118/87000-MS.

Lejon, K., Thingvoll, J. T., Vollen, E. A. et al. 1995. Formation Damage Due to Losses of Ca-Based Brine and How It Was Revealed Through Post Evaluation of Scale Dissolver and Scale Inhibitor Squeeze Treatments. Paper presented at the SPE European Formation Damage Conference, The Hague, The Netherlands, 15–16 May. https://doi.org/10.2118/30086-MS.

Li, Y.-H., Crane, S. D., Scott, E. M. et al. 1996. Waterflood Geochemical Modeling and a Prudhoe Bay Zone 4 Case Study. Paper presented at the SPE Formation Damage Control Symposium, Lafayette, Louisiana, USA, 14–15 February. SPE-31136-MS. https://doi.org/10.2118/31136-MS.

Ligthelm, D. J., de Boer, R. B., Brint, J. F. et al. 1991. Reservoir Souring: An Analytical Model for H_2S Generation and Transportation in an Oil Reservoir Owing to Bacterial Activity. Paper presented at Offshore Europe, Aberdeen, UK, 3–6 September. SPE-23141-MS. https://doi.org/10.2118/23141-MS.

Ly, K. T., Blumer, D. J., Bohon, W. M. et al. 1998. Novel Chemical Dispersant for Removal of Organic/Inorganic "Schmoo" Scale in Produced Water Injection Systems. Paper presented at CORROSION 98, San Diego, California, USA, 22–27 March. NACE-98073.

Maas, J. G., Wit, K., and Morrow, N. R. 2001. Enhanced Oil Recovery by Dilution of Injection Brine: Further Interpretation of Experimental Results. SCA 2001-13.

Mackay, E. J. 2002. Modelling of In-Situ Scale Deposition: The Impact of Reservoir and Well Geometries and Kinetic Reaction Rates. Paper presented at the International Symposium on Oilfield Scale, Aberdeen, UK, 30–31 January. SPE-74683-MS. https://doi.org/10.2118/74683-MS.

Mackay, E. J. and Jordan, M. M. 2003. SQUEEZE Modelling: Treatment Design and Case Histories. Paper presented at the SPE European Formation Damage Conference, The Hague, The Netherlands, 13–14 May. SPE-82227-MS. https://doi.org/10.2118/82227-MS.

Mackay, E. J., Jordan, M. M., Feasey, N. D. et al. 2005a. Integrated Risk Analysis for Scale Management in Deepwater Developments. *SPE Prod & Fac* 20 (2): 138–154. SPE-87459-PA. https://doi.org/10.2118/87459-PA.

Mackay, E. J., Jordan, M. M., and Torabi, F. 2002. Predicting Brine Mixing Deep Within the Reservoir, and the Impact on Scale Control in Marginal and Deepwater Developments. Paper presented at the International Symposium and Exhibition on Formation Damage Control, Lafayette, Louisiana, USA, 20–21 February. SPE-73779-MS. https://doi.org/10.2118/73779-MS.

Mackay, E. J., Jordan, M. M., and Torabi, F. 2003. Predicting Brine Mixing Deep Within the Reservoir and Its Impact on Scale Control in Marginal and Deepwater

Developments. *SPE Prod & Fac* **18** (3): 210–220. SPE-85104-PA. https://doi.org/10.2118/85104-PA.

Mackay, E. J., Sorbie, K. S., Boak, L. S. et al. 2005b. What Level of Sulphate Reduction Is Required to Eliminate the Need for Scale Inhibitor Squeezing? Paper presented at the SPE International Symposium on Oilfield Scale, Aberdeen, UK, 11–12 May. SPE-95089-MS. https://doi.org/10.2118/95089-MS.

Mackay, E. J., Sorbie, K. S., Kavle, V. M. et al. 2006. Impact of In Situ Sulphate Stripping on Scale Management in the Gyda Field. Paper presented at the SPE International Oilfield Scale Symposium, Aberdeen, UK, 31 May–1 June. SPE-100516-MS. https://doi.org/10.2118/100516-MS.

Macleod, N., Bryan, T., Buckley, A. J. et al. 1994. A Novel Biocide for Oilfield Applications. SPE-30171-MS.

Martins, J. P., Murray, L. R., Clifford, P. J. et al. 1995. Produced-Water Reinjection and Fracturing in Prudhoe Bay. *SPE Res Eng* **10** (3): 176–182. SPE-28936-PA. https://doi.org/10.2118/28936 PA.

Maudgalya, S., Knapp, R. M., and McInerney, M. 2007. Microbially Enhanced Oil Recovery Technologies: A Review of the Past, Present and Future. Paper presented at the Production and Operations Symposium, Oklahoma City, Oklahoma, USA, 31 March–3 April. SPE-106978-MS. https://doi.org/10.2118/106978-MS.

Maxwell, S. and Spark, I. 2005. Souring of Reservoirs by Bacterial Activity During Seawater Flooding. Paper presented at the SPE International Symposium on Oilfield Chemistry, The Woodlands, Texas, USA, 2–4 February. SPE-93231-MS. https://doi.org/10.2118/93231-MS.

McCartney, D. M. and Oleszkiewicz, J. A. 1991. Sulfide Inhibition of Anaerobic Degradation of Lactate and Acetate. *Water Research* **25** (2): 203–209. https://doi.org/10.1016/0043-1354(91)90030-T.

McElhiney, J. E., Tomson, M. B., and Kan, A. T. 2006. Design of Low Sulphate Seawater Injection Based Upon Kinetic Limits. Paper presented at the SPE International Oilfield Scale Symposium, Aberdeen, UK, 31 May–1 June. SPE-100480-MS. https://doi.org/10.2118/100480-MS.

McGuire, P. L., Chatham, J. R., Paskvan, F. K. et al. 2005. Low Salinity Oil Recovery: An Exciting New EOR Opportunity for Alaska's North Slope. Paper presented at the SPE Western Regional Meeting, Irvine, California, USA, 30 March–1 April. SPE-93903-MS. https://doi.org/10.2118/93903-MS.

Miller, S. G. and Robuck, R. D. 1972. The Stretford Process at East Wilmington Field. *J Pet Technol* **24** (5): 545–548. SPE-3700-PA. https://doi.org/10.2118/3700-PA.

Mitchell, A. F., Skjevrak, I., and Waage, J. 2017. A Re-Evaluation of Reservoir Souring Patterns and Effect of Mitigation in a Mature North Sea Field. Paper presented at the SPE International Conference on Oilfield Chemistry, Montgomery, Texas, USA, 3–5 April. SPE-3700-PA. https://doi.org/10.2118/184587-MS.

Muecke, T. W. 1979. Formation Fines and Factors Controlling Their Movement in Porous Media. *J Pet Technol* **31** (2): 144–150. SPE-7007-PA. https://doi.org/10.2118/7007-PA.

Muyzer, G. and Stams, A. J. M. 2008. The Ecology and Biotechnology of Sulphate-Reducing Bacteria. *Nature Reviews Microbiology* **6** (6): 441–454. https://doi.org/10.1038/nrmicro1892.

Nasr-El-Din, H. A. and Al-Humaidan, A. Y. 2001. Iron Sulfide Scale: Formation, Removal and Prevention. Paper presented at the International Symposium on

Oilfield Scale, Aberdeen, UK, 30–31 January. SPE-68315-MS. https://doi.org/10.2118/68315-MS.

Norman, C. A. and Smith, J. E. 2000. Experience Gained from 318 Injection Well KOH Clay Stabilization Treatments. Paper presented at the SPE Rocky Mountain Regional/Low-Permeability Reservoirs Symposium and Exhibition, Denver, Colorado, USA, 12–15 March. SPE-60307-MS. https://doi.org/10.2118/60307-MS.

Oduola, L., Igwebueze, C., Dede, A. et al. 2009. Reservoir Souring Mitigation in the Bonga West Africa Deepwater Field Using Calcium Nitrate. Paper presented at the BIPOG-3 International Conference on Biotechnology for Improved Production of Oil and Gas in the Gulf of Guinea.

Okpala, G. N. and Voordouw, G. 2018. Comparison of Nitrate and Perchlorate in Controlling Sulfidogenesis in Heavy Oil-Containing Bioreactors. *Frontiers in Microbiology*. https://doi.org/10.3389/fmicb.2018.02423.

Østvold, T. and Randhol, P. 2001. Kinetics of $CaCO_3$ Scale Formation. The Influence of Temperature, Supersaturation and Ionic Composition. Paper presented at the International Symposium on Oilfield Scale, Aberdeen, UK, 30–31 January. SPE-68302-MS. https://doi.org/10.2118/68302-MS.

Owens, W. W. and Archer, D. L. 1971. The Effect of Rock Wettability on Oil-Water Relative Permeability Relationships. *J Pet Technol* 23 (7): 873–878. SPE-3034-PA. https://doi.org/10.2118/3034-PA.

Pedenaud, P., Hurtevent, C., and Baraka-Lokmane, S. 2012. Industrial Experience in Sea Water Desulfation. Paper presented at the SPE International Conference on Oilfield Scale, Aberdeen, UK, 30–31 May. SPE-155123-MS. https://doi.org/10.2118/155123-MS.

Peters, F. W. and Stout, C. M. 1977. Clay Stabilization During Fracturing Treatments with Hydrolyzable Zirconium Salts. *J Pet Technol* 29 (2): 187–194. SPE-5687-PA. https://doi.org/10.2118/5687-PA.

Poggesi, G., Hurtevent, C., and Brazy, J. L. 2001. Scale Inhibitor Injection via the Gas Lift System in High Temperature Block 3 Fields in Angola. Paper presented at the International Symposium on Oilfield Scale, Aberdeen, UK, 30–31 January. SPE-68301-MS. https://doi.org/10.2118/68301-MS.

Postgate, J. R. 1984. *The Sulphate-Reducing Bacteria*. Cambridge, UK: Cambridge University Press.

Pritchard, A. M., Birch, W., and Brabon, S. 2000. Assessing the Effect of the Application of Magnetic Treatment for the Reduction of Oilfield Scale. Paper presented at CORROSION 2000, Orlando, Florida, USA, 26–31 March. NACE-00110.

Pruess, K., Xu, T., Birkle, P. et al. 2006. Using Laboratory Flow Experiments and Reactive Chemical Transport Modeling for Designing Waterflooding of the Agua Fr'a Reservoir, Poza Rica-Altamira Field, Mexico. Paper presented at the International Oil Conference and Exhibition in Mexico, Cancun, Mexico, 31 August–2 September. SPE-103869-MS. https://doi.org/10.2118/103869-MS.

Przybylinski, J. L. 2001. Iron Sulfide Scale Deposit Formation and Prevention Under Anaerobic Conditions Typically Found in the Oil Field. Paper presented at the SPE International Symposium on Oilfield Chemistry, Houston, Texas, USA, 13–16 February. SPE-65030-MS. https://doi.org/10.2118/65030-MS.

Pu, H., Xie, X., Yin, P. et al. 2010. Low-Salinity Waterflooding and Mineral Dissolution. Paper presented at the SPE Annual Technical Conference and Exhibition, Florence, Italy, 19–22 September. SPE-134042-MS. https://doi.org/10.2118/134042-MS.

Putnis, A., Putnis, C. V., and Paul, J. M. 1995. The Efficiency of a DTPA-Based Solvent in the Dissolution of Barium Sulfate Scale Deposits. Paper presented at the SPE International Symposium on Oilfield Chemistry, San Antonio, Texas, USA, 14–17 February. SPE-29094-MS. https://doi.org/10.2118/29094-MS.

Rabaioli, M. R. and Lockhart, T. P. 1995. Solubility and Phase Behaviour of Polyacrylate Scale Inhibitors and Their Implications for Precipitation Squeeze Treatment. Paper presented at the SPE International Symposium on Oilfield Chemistry, San Antonio, Texas, USA, 14–17 February. SPE-28998-MS. https://doi.org/10.2118/28998-MS.

Ramstad, K., Rohde, H. C., Tydal, T. et al. 2009. Scale Squeeze Evaluation Through Improved Sample Preservation, Inhibitor Detection and Minimum Inhibitor Concentration Monitoring. *SPE Prod & Oper* **24** (4): 530–542. SPE-114085-PA. https://doi.org/10.2118/114085-PA.

Rassenfoss, S. 2011. From Bacteria to Barrels: Microbiology Having an Impact on Oil Fields. *J Pet Technol* **63** (11): 32–38. SPE-1111-0032-JPT. https://doi.org/10.2118/1111-0032-JPT.

Ravenscroft, P. D., Cowie, L. G., and Smith, P. S. 1996. Magnus Scale Inhibitor Squeeze Treatments – A Case History. Paper presented at the SPE Annual Technical Conference and Exhibition, Denver, Colorado, USA, 6–9 October. SPE-36612-MS. https://doi.org/10.2118/36612-MS.

Reed, M. G. 1972. Stabilization of Formation Clays with Hydroxy-Aluminum Solutions. *J Pet Technol* **24** (7): 860–864. SPE-3694-PA. https://doi.org/10.2118/3694-PA.

Reis, M. A. M., Almeida, J. S., Lemos. P. C. et al. 1992. Effect of Hydrogen Sulfide on Growth of Sulfate Reducing Bacteria. *Biotechnology and Bioengineering* **40** (5): 593–600. https://doi.org/10.1002/bit.260400506.

Robbana, E., Buikema, T. A., Mair, C. et al. 2012. Low Salinity Enhanced Oil Recovery – Laboratory to Day One Field Implementation - LoSal EOR into the Clair Ridge Project. Paper presented at the Abu Dhabi International Petroleum Conference and Exhibition, Abu Dhabi, UAE, 11–14 November. SPE-161750-MS. https://doi.org/10.2118/161750-MS.

Robertson, E. P. 2007. Low-Salinity Waterflooding to Improve Oil Recovery—Historical Field Evidence. Paper presented at the SPE Annual Technical Conference and Exhibition, Anaheim, California, USA, 11–14 November. SPE-109965-MS. https://doi.org/10.2118/109965-MS.

Robinson, K., Ginty, W. R., Samuelsen, E. H. et al. 2010. Reservoir Souring in a Field with Sulphate Removal: A Case Study. Paper presented at the SPE Annual Technical Conference and Exhibition, Florence, Italy, 19–22 September. SPE-132697-MS. https://doi.org/10.2118/132697-MS.

Rodrigues, V. F., Frota, H. M., Loures, L. G. L. et al. 2009. Hydrate Blockage in Subsea Water Injection Wells – Causes and Preventive Procedures. Paper presented at the Society of Petroleum Engineers 8th European Formation Damage Conference, Scheveningen, The Netherlands, 27–29 May. SPE-120273-MS. https://doi.org/10.2118/120273-MS.

Rogerson, N. and Laing, T. J. A. 2007. Subsea Raw Seawater Injection System—A World First. Paper presented at Offshore Europe, Aberdeen, Scotland, UK, 4–7 September. SPE-109090-MS. https://doi.org/10.2118/109090-MS.

Salma, T. 2000. Cost Effective Removal of Iron Sulfide and Hydrogen Sulfide from Water Using Acrolein. Paper presented at the SPE Permian Basin Oil and Gas

Recovery Conference, Midland, Texas, USA, 21–23 March. SPE-59708-MS. https://doi.org/10.2118/59708-MS.

Scheuerman, R. F. and Bergersen, B. M. 1990. Injection-Water Salinity, Formation Pretreatment, and Well-Operations Fluid-Selection Guidelines. *J Pet Technol* **42** (7): 836–845. SPE-18461-PA. https://doi.org/10.2118/18461-PA.

Schofield, M. J., Waterton, K. C., Evans, T. et al. 2004. Corrosion Behavior of Carbon Steel, Low Alloy Steel and CRA's in Partially Deaerated Sea Water and Commingled Produced Water. Paper presented at CORROSION 2004, New Orleans, Louisiana, USA, 28 March–1 April. NACE-04139.

Seccombe, J., Lager, A., Jerauld, G. et al. 2010. Demonstration of Low-Salinity EOR at Interwell Scale, Endicott Field, Alaska. Paper presented at the SPE Improved Oil Recovery Symposium, Tulsa, Oklahoma, USA, 24–28 April. SPE-129692-MS. https://doi.org/10.2118/129692-MS.

Seccombe, J. C., Lager, A., Webb, K. J. et al. 2008. Improving Wateflood Recovery: LoSal™ EOR Field Evaluation. Paper presented at the SPE Symposium on Improved Oil Recovery, Tulsa, Oklahoma, USA, 20–23 April. SPE-113480-MS. https://doi.org/10.2118/113480-MS.

Selle, O. M., Haavind, F., Haukland, M. H. et al. 2010. Downhole Scale Control on Heidrun Field Using Scale Inhibitor Impregnated Gravel. Paper presented at the SPE International Conference on Oilfield Scale, Aberdeen, UK, 26–27 May. SPE-130788-MS. https://doi.org/10.2118/130788-MS.

Seto, C. J. and Beliveau, D. A. 2000. Reservoir Souring in the Caroline Field. Paper presented at the SPE/CERI Gas Technology Symposium, Calgary, Alberta, Canada, 3–5 April. SPE-59778-MS. https://doi.org/10.2118/59778-MS.

Shuler, P. J., Baudoin, D. A., and Weintritt, D. J. 1995. Diagnosis and Prevention of NORM at Eugene Island 341-A. Paper presented at the SPE/EPA Exploration and Production Environmental Conference, Houston, Texas, USA, 27–29 March. SPE-29711-MS. https://doi.org/10.2118/29711-MS.

Simpson, C., Graham, G. M., Collins, I. R. et al. 2005. Sulphate Removal for Barium Sulphate Mitigation – Kinetic vs. Thermodynamic Controls in Mildly Oversaturated Conditions. Paper presented at the SPE International Symposium on Oilfield Scale, Aberdeen, UK, 11–12 May. SPE-95082-MS. https://doi.org/10.2118/95082-MS.

Skrettingland, K., Holt, T., Tweheyo, M. T. et al. 2010. Snorre Low Salinity Water Injection – Core Flooding Experiments and Single Well Field Pilot. Paper presented at the SPE Improved Oil Recovery Symposium, Tulsa, Oklahoma, USA, 24–28 April. SPE-129877-MS. https://doi.org/10.2118/129877-MS.

Smith, P. S., Clement, C. C., and Rojas, A. M. 2000. Combined Scale Removal and Scale Inhibition Treatments. Paper presented at the International Symposium on Oilfield Scale, Aberdeen, UK, 26–27 January. SPE-60222-MS. https://doi.org/10.2118/60222-MS.

Sorbie, K. S. and Laing, N. 2004. How Scale Inhibitors Work: Mechanisms of Selected Barium Sulphate Scale Inhibitors Across a Wide Temperature Range. Paper presented at the SPE International Symposium on Oilfield Scale, Aberdeen, UK, 26–27 May. SPE-87470-MS. https://doi.org/10.2118/87470-MS.

Stott, J. F. D. 2012. Implementation of Nitrate Treatment for Reservoir Souring Control: Complexities and Pitfalls. Paper presented at the SPE International Conference & Workshop on Oilfield Corrosion, Aberdeen, UK, 28–29 May. SPE-155155-MS. https://doi.org/10.2118/155155-MS.

Strand, S., Austad, T., Puntervold, T. et al. 2008. "Smart Water" for Oil Recovery from Fractured Limestone: A Preliminary Study. *Energy Fuels* **22** (5): 3126–3133. https://doi.org/10.1021/ef800062n.

Sunde, E., Beeder, J., Nilsen, R. K. et al. 1992. Aerobic Microbial Enhanced Oil Recovery for Offshore Use. Paper presented at the SPE/DOE Enhanced Oil Recovery Symposium, Tulsa, Oklahoma, USA, 22–24 April. SPE-24204-MS. https://doi.org/10.2118/24204-MS.

Sunde, E., Bodtker, G., Lillebo, B.-L. et al. 2004. H₂S Inhibition by Nitrate Injection on the Gullfaks Field. Paper presented at CORROSION 2004, New Orleans, Louisiana, USA, 28 March–1 April. NACE-04760.

Sunde, E., Thorstenson, T., Torsvik, T. et al. 1993. Field-Related Mathematical Model to Predict and Reduce Reservoir Souring. Paper presented at the SPE International Symposium on Oilfield Chemistry, New Orleans, Louisiana, USA, 2–5 March. SPE-25197-MS. https://doi.org/10.2118/25197-MS.

Sutherland, L. and Jordan, M. 2016. Enhancing Scale Inhibitor Squeeze Retention with Additives. Paper presented at the SPE International Oilfield Scale Conference and Exhibition, Aberdeen, Scotland, UK, 11–12 May. SPE-179888-MS. https://doi.org/10.2118/179888-MS.

Talbot, R. E., Gilbert, P. D., Veale, M. A. et al. 2002. TetrakisHydroxymethylPhosphonium Sulfate (THPS) for Dissolving Iron Sulfides Downhole and Topsides – A Study of the Chemistry Influencing Dissolution. Paper presented at CORROSION 2002, Denver, Colorado, USA, 7–11 April. NACE-02030.

Tang, G. Q. and Morrow, N. R. 1997. Salinity, Temperature, Oil Composition, and Oil Recovery by Waterflooding. *SPE Res Eng* **12** (4): 269–276. SPE-36680-PA. https://doi.org/10.2118/36680-PA.

Tang, G. and Morrow, N. R. 1999a. Oil Recovery by Waterflooding and Imbibition – Invading Brine Cation Valency and Salinity. SCA-9911.

Tang, G. Q. and Morrow, N. R. 1999b. Influence of Brine Composition and Fines Migration on Crude Oil/Brine/Rock Interactions and Oil Recovery. *Journal of Petroleum Science and Engineering* **24** (2–4): 99–111. https://doi.org/10.1016/S0920-4105(99)00034-0.

Telang, A. J., Ebert, S., Foght, J. M. et al. 1997. Effect of Nitrate Injection on the Microbial Community in an Oil Field as Monitored by Reverse Sample Genome Probing. *Applied and Environmental Microbiology* **63** (5): 1785–1793.

Templeton, C. C. 1960. Solubility of Barium Sulfate in Sodium Chloride Solutions from 25°C to 95°C. *J. Chem. Eng. Data* **5** (4): 514–516. https://doi.org/10.1021/je60008a028.

Thorstenson, T., Sunde, E., Bodtker, G. et al. 2002. Biocide Replacement by Nitrate in Sea Water Injection Systems. Paper presented at CORROSION 2002, Denver, Colorado, USA, 7–11 April. NACE-02033.

Tjomsland, T., Grotle, M. N., and Vikane, O. 2001. Scale Control Strategy and Economical Consequences of Scale at Veslefrikk. Paper presented at the International Symposium on Oilfield Scale, Aberdeen, UK, 30–31 January. SPE-68308-MS. https://doi.org/10.2118/68308-MS.

Tomson, M. B., Fu, G., Watson, M. A. et al. 2003. Mechanisms of Mineral Scale Inhibition. *SPE Prod & Fac* **18** (3): 192–199. SPE-84958-PA. https://doi.org/10.2118/84958-PA.

Town, K., Sheehy, A., and Govreau, B. R. 2009. MEOR Success in Southern Saskatchewan. Paper presented at the SPE Annual Technical Conference and

Exhibition, New Orleans, Louisiana, USA, 4–7 October. SPE-124319-MS. https://doi.org/10.2118/124319-MS.

Vazquez, O., Mackay, E. J., Jordan, M. M. et al. 2009. Impact of Mutual Solvent Preflushes on Scale Squeeze Treatments: Extended Squeeze Lifetime and Improved Well Clean-up Time. Paper presented at the 8th European Formation Damage Conference, Scheveningen, The Netherlands, 27–29 May. SPE-121857-MS. https://doi.org/10.2118/121857-MS.

Vazquez, O., Mackay, E., Tjomsland, T. et al. 2014. Use of Tracers to Evaluate and Optimize Scale-Squeeze-Treatment Design in the Norne Field. *SPE Prod & Oper* **29** (1): 5–13. SPE-164114-PA. https://doi.org/10.2118/164114-PA.

Vazquez, O., McCartney, R., and Mackay, E. 2013. Produced Water Chemistry History Matching Using a 1D Reactive Injector Producer Reservoir Model. Paper presented at the SPE International Symposium on Oilfield Chemistry, The Woodlands, Texas, USA, 8–10 April. SPE-164113-MS. https://doi.org/10.2118/164113 MS.

Venzlaff, H., Enning, D., Srinivasan, J. et al. 2013. Accelerated Cathodic Reaction in Microbial Corrosion of Iron Due to Direct Electron Uptake by Sulfate-Reducing Bacteria. *Corrosion Science* **66**: 88–96. https://doi.org/10.1016/j.corsci.2012.09.006.

Vetter, O. J. and Farone, W. A. 1987. Calcium Carbonate Scale in Oilfield Operations. Paper presented at the SPE Annual Technical Conference and Exhibition, Dallas, Texas, USA, 27–30 September. SPE-16908-MS. https://doi.org/10.2118/16908-MS.

Vik, E. A. 2008. PWRI – An Environmental Friendly, but Challenging Water Management Option. Experiences from the NCS and Expected Challenges for the Barents Sea and Lofoten Area. Paper presented at the 5th TEKNA International Conference on PW Management, Stavanger, Norway, 22–23 January.

Vik, E. A., Janbu, A. O., Garshol, F. K. et al. 2007. Nitrate Based Souring Mitigation of Produced Water – Side Effects and Challenges from the Draugen Produced Water Re-Injection Pilot. Paper presented at the International Symposium on Oilfield Chemistry, Houston, Texas, USA, 28 February–2 March. SPE-106178-MS. https://doi.org/10.2118/106178-MS.

Vledder, P., Gonzalez, I. E., Carrera Fonseca, J. C. et al. 2010. Low Salinity Water Flooding: Proof of Wettability Alteration on a Field Wide Scale. Paper presented at the SPE Improved Oil Recovery Symposium, Tulsa, Oklahoma, USA, 24–28 April. SPE-129564-MS. https://doi.org/10.2118/129564-MS.

Voordouw, G., Agrawal, A., Park, H.-S. et al. 2011. Souring Treatment with Nitrate in Fields from Which Oil Is Produced by Produced Water Reinjection. Paper presented at the SPE International Symposium on Oilfield Chemistry, The Woodlands, Texas, USA, 11–13 April. SPE-141354-MS. https://doi.org/10.2118/141354-MS.

Voordouw, G., Buziak, B., Lin, S. et al. 2007. Use of Nitrate or Nitrite for the Management of the Sulfur Cycle in Oil and Gas Fields. Paper presented at the International Symposium on Oilfield Chemistry, Houston, Texas, USA, 28 February–2 March. SPE-106288-MS. https://doi.org/10.2118/106288-MS.

Vu, V. K., Hurtevent, C., and Davis, R. A. 2000. Eliminating the Need for Scale Inhibition Treatments for Elf Exploration Angola's Girassol Field. Paper presented at the International Symposium on Oilfield Scale, Aberdeen, UK, 26–27 January. SPE-60220-MS. https://doi.org/10.2118/60220-MS.

Waite, R. J., Eden, R., and Cousins, A. R. 1996. Seabed Rawwater Injection: An Alternative Pressure Maintenance System. Paper presented at the SPE Annual Technical Conference and Exhibition, Denver, Colorado, USA, 6–9 October. SPE-36677-MS. https://doi.org/10.2118/36677-MS.

Wang, S. and Civan, F. 2005. Preventing Asphaltene Deposition in Oil Reservoirs by Early Water Injection. Paper presented at the SPE Production Operations Symposium, Oklahoma City, Oklahoma, USA, 16–19 April. SPE-94268-MS. https://doi.org/10.2118/94268-MS.

Warner, H. R., Jr. 2015. *The Reservoir Engineering Aspects of Waterflooding*, second edition, Vol. 3. Richardson, Texas: Monograph Series, Society of Petroleum Engineers.

Wat, R., Montgomerie, H., Hagen, T. et al. 1999. Development of an Oil-Soluble Scale Inhibitor for a Subsea Satellite Field. Paper presented at the SPE International Symposium on Oilfield Chemistry, Houston, Texas, USA, 16–19 February. SPE-50706-MS. https://doi.org/10.2118/50706-MS.

Weaver, J. D., Nguyen, P. D., and Loghry, R. 2011. Stabilizing Fracture Faces in Water-Sensitive Shale Formations. Paper presented at the SPE Eastern Regional Meeting, Columbus, Ohio, USA, 17–19 August. SPE-149218-MS. https://doi.org/10.2118/149218-MS.

Webb, K. J., Black, C. J. J., and Al-Ajeel, H. 2004. Low Salinity Oil Recovery – Log-Inject-Log. Paper presented at the SPE/DOE Symposium on Improved Oil Recovery, Tulsa, Oklahoma, USA, 17–21 April. SPE-89379-MS. https://doi.org/10.2118/89379-MS.

Webb, K. J., Black, C. J. J., and Tjetland, G. 2005. A Laboratory Study Investigating Methods for Improving Oil Recovery in Carbonates. Paper presented at the International Petroleum Technology Conference, Doha, Qatar, 21–23 November. IPTC-10506-MS. https://doi.org/10.2523/IPTC-10506-MS.

Willhite, G. P. 1986. *Waterflooding*, Vol. 3. Richardson, Texas: Textbook Series, Society of Petroleum Engineers.

Wylde, J. J. 2014. Sulfide Scale Control in Produced Water Handling and Injection Systems: Best Practices and Global Experience Overview. Paper presented at the SPE International Oilfield Scale Conference and Exhibition, Aberdeen, Scotland, 14–15 May. SPE-169776-MS. https://doi.org/10.2118/169776-MS.

Wylde, J. J. and Fell, D. 2008. Scale Inhibitor Solutions for High Temperature ESP Lifted Wells in Northern California: A Case History of Failure Followed by Success. Paper presented at the SPE International Oilfield Scale Conference, Aberdeen, UK, 28–29 May. SPE-113826-MS. https://doi.org/10.2118/113826-MS.

Yap, J., Fuller, M. J., Schafer, L. et al. 2010. Removing Iron Sulfide Scale: A Novel Approach. Paper presented at the Abu Dhabi International Petroleum Exhibition and Conference, Abu Dhabi, UAE, 1–4 November. SPE-138520-MS. https://doi.org/10.2118/138520-MS.

Yaseen, S. and Mansoori, G. A. 2018. Asphaltene Aggregation Due to Waterflooding (A Molecular Dynamics Study). *Journal of Petroleum Science and Engineering* 170: 177–183. https://doi.org/10.1016/j.petrol.2018.06.043.

Yousef, A. A., Al-Saleh, S., and Al-Jawfi, M. S. 2012a. Improved/Enhanced Oil Recovery from Carbonate Reservoirs by Tuning Injection Water Salinity and Ionic Content. Paper presented at the SPE Improved Oil Recovery Symposium, Tulsa, Oklahoma, USA, 14–18 April. SPE-154076-MS. https://doi.org/10.2118/154076-MS.

Yousef, A. A., Al-Saleh, S., and Al-Jawfi, M. S. 2012b. The Impact of the Injection Water Chemistry on Oil Recovery from Carbonate Reservoirs. Paper presented at the SPE EOR Conference at Oil and Gas West Asia, Muscat, Oman, 16–18 April. SPE-154077-MS. https://doi.org/10.2118/154077-MS.

Yousef, A. A., Al-Saleh, S. H., Al-Kaabi, A. et al. 2011. Laboratory Investigation of the Impact of Injection-Water Salinity and Ionic Content on Oil Recovery from Carbonate Reservoirs. *SPE Res Eval & Eng* **14** (5): 578–593. SPE-137634-PA. https://doi.org/10.2118/137634-PA.

Yuan, M. 2001. Barium Sulfate Scale Inhibition in the Deepwater Cold Temperature Environment. Paper presented at the International Symposium on Oilfield Scale, Aberdeen, UK, 30–31 January. SPE-68311-MS. https://doi.org/10.2118/68311-MS.

Zahid, A., Shapiro, A. A., and Skauge, A. 2012. Experimental Studies of Low Salinity Water Flooding Carbonate: A New Promising Approach. Paper presented at the SPE EOR Conference at Oil and Gas West Asia, Muscat, Oman, 16–18 April. SPE-155625-MS. https://doi.org/10.2118/155625-MS.

Zahner, R. L., Govreau, B. R., and Sheehy, A. 2010. MEOR Success in Southern California. Paper presented at the SPE Improved Oil Recovery Symposium, Tulsa, Oklahoma, USA, 24–28 April. SPE-129742-MS. https://doi.org/10.2118/129742-MS.

Zahner, R. L., Tapper, S., Marcotte, B. W. G. et al. 2011. What Has Been Learned from A Hundred MEOR Applications. Paper presented at the SPE Enhanced Oil Recovery Conference, Kuala Lumpur, Malaysia, 19–21 July. SPE-145054-MS. https://doi.org/10.2118/145054-MS.

Zhou, Z. J., Gunter, W. O., and Jonasson, R. G. 1995. Controlling Formation Damage Using Clay Stabilizers: A Review. Paper presented at the Petroleum Society of Canada Annual Technical Meeting, Calgary, Alberta, 7–9 June. PETSOC-95-71. https://doi.org/10.2118/95-71.

Zhu, F., Kuijvenhoven, C., Borhan, N. B. et al. 2016. Learnings from Reservoir Souring Assessment on an Offshore Malaysian Field. Paper presented at the SPE EOR Conference at Oil and Gas West Asia, Muscat, Oman, 21–23 March. SPE-179787-MS. https://doi.org/10.2118/179787-MS.

SI Metric Conversion Factors

bbl × 1.589 873 E–01 = m^3

°F (°F – 32)/1.8 = °C

ft × 3.048* E–01 = m

lbm × 0.453592 = kg

psi × 6.895 = kPa

*Conversion factor is exact.

PetroBriefs

SPE's newest book series is meant to quickly bring each reader up to speed on an emerging technology or specialized topic. At approximately 100 pages, most PetroBriefs are available in both softcover and popular eBook formats.

To see a current list of available PetroBriefs, visit store.spe.org.

Authored by Dave Chappell

Waterflooding: Chemistry
Waterflooding: Facilities and Operations
Waterflooding: Design and Development
Waterflooding: Surveillance and Remediation
Waterflooding: Injection Regime and Injection Wells

SPE International

Society of Petroleum Engineers

SPE is the recognized leader for publications in the upstream oil and gas industry, and the SPE Bookstore is your source for the books that set the standards of excellence.

Don't miss out on the latest from SPE. Sign up to receive information about new releases at store.spe.org.